Power Systems

Electrical power has been the technological foundation of industrial societies for many years. Although the systems designed to provide and apply electrical energy have reached a high degree of maturity, unforeseen problems are constantly encountered, necessitating the design of more efficient and reliable systems based on novel technologies. The book series Power Systems is aimed at providing detailed, accurate and sound technical information about these new developments in electrical power engineering. It includes topics on power generation, storage and transmission as well as electrical machines. The monographs and advanced textbooks in this series address researchers, lecturers, industrial engineers and senior students in electrical engineering.

Power Systems is indexed in Scopus

More information about this series at http://www.springer.com/series/4622

Mehdi Rahmani-Andebili
Editor

Design, Control, and Operation of Microgrids in Smart Grids

Springer

Editor
Mehdi Rahmani-Andebili
Department of Engineering Technology
State University of New York, Buffalo State
Buffalo, NY, USA

ISSN 1612-1287 ISSN 1860-4676 (electronic)
Power Systems
ISBN 978-3-030-64630-1 ISBN 978-3-030-64631-8 (eBook)
https://doi.org/10.1007/978-3-030-64631-8

This Springer imprint is published by the registered company Springer Nature Switzerland AG
The registered company address is: Gewerbestrasse 11, 6330 Cham, Switzerland

Preface

A microgrid is a group of interconnected loads and distributed energy resources that can act as a single controllable entity and connect and disconnect from the grid to operate in a grid-connected or island mode. Microgrids have received considerable attention in the recent years as they can mitigate the growing problems of power system.

This book covers the recent research advancements in the design, control, and operation of microgrids in a smart grid infrastructure. The first chapter deals with the design and control of a single input interval type-2 fuzzy logic controller for the frequency regulation of a multi-microgrids system. Herein, the presence of electric vehicles and vehicle-to-grid technology are considered.

In the second chapter, a robust distributed noise-resilient control for the DC microgrids is proposed. Herein, the dynamic communication topology and multiplicative noise disturbances are taken into account to guarantee the stability operation of the closed-loop system through adopting stochastic stability theory and Lyapunov function.

The third chapter presents a dynamic design framework of microgrids to provide the optimal capacity of the distributed energy resources and the optimal investment timing for the stakeholders. Herein, the short-term and long-term uncertainty issues of the problem parameters are considered, and a mixed integer multi-objective linear programming approach is applied.

The fourth chapter presents a bi-level distributed operation approach for the interconnected microgrid system to optimally coordinate the operation of the microgrids owned by different owners while considering the power market uncertainties. In this chapter, a chance constraint programming and a combination of the analytical target cascading and augmented Lagrange method are applied in the lower and upper levels.

The fifth chapter develops a two-level optimization model for the microgrid bidding schemes in a multi-agent system to promote the interaction between the main grid and the microgrid. Herein, an artificial immune system is applied as the optimization technique.

The sixth chapter is concerned with the microgrid formation, which is dividing a microgrid into some sub-microgrids to mitigate the issues caused by natural disasters and cyber-physical attacks. Herein, the microgrid formation is carried out by switching the candidate breakers and tie-line switches to increase the resilience and load restoration of the microgrid.

Buffalo, NY, USA Mehdi Rahmani-Andebili

Contents

Chapter 1
Multi-microgrids with a Frequency Regulation-Based V2G Technology: Systems Analysis, Modeling, and Control

Meysam Gheisarnejad and Mohammad-Hassan Khooban

Abstract This chapter presents novel structured single-input interval type-2 fuzzy logic controllers (SI-IT2-FLCs) for the frequency damping of multi-microgrids (MMGs), whereas the application of electric vehicles (EVs) is considered in this context. For this purpose, a new SI-IT2-fuzzy PD/fuzzy PI (SI-IT2-FPD/FPI) controller is designed on two levels. Initially, an improved whale optimization algorithm, called IWOA, is adopted to adjust the setting of the gains embedded in the FPD/FPI section effectively. Then, the impact of the footprint of uncertainty (FOU), to offer extra design freedom, on control surface generation of SI-IT2-FLC has been investigated. In this way, various control surfaces were generated by varying a single coefficient which forms the FOU. Lastly, by adopting hardware-in-the-loop (HIL) simulator, the feasibility and usefulness of the suggested framework are verified from a real-time perspective.

Keywords Secondary load frequency control (LFC) · Multi-microgrid (MMG) · Footprint of uncertainty (FOU)) · Electric vehicle (EV) · Hardware-in-the-loop (HIL)

1.1 Introduction

Over the last decades, the environmental issues related to fossil fuels, along with the inevitable rise in energy prices, led to a new tendency of generating power locally by the integration of vast contributions of renewable energy resources (RESs) [1–4]. The RESs, such as solar and wind energy, are abundant in nature; as a result, deep penetration of these resources will reduce environmental pollution and cover other conventional power supply constraints such as security and reliability. The

M. Gheisarnejad
Department of Electrical Engineering, Islamic Azad University, Najafabad Branch, Isfahan, Iran

M.-H. Khooban (✉)
Department of Electrical and Computer Engineering, Aarhus University, Aarhus, Denmark
e-mail: khooban@eng.au.dk

M. Rahmani-Andebili (ed.), *Design, Control, and Operation of Microgrids in Smart Grids*, Power Systems, https://doi.org/10.1007/978-3-030-64631-8_1

intermittent nature of RESs leads to power outages or power surges. To deal with this unpleasant property and to provide reliable operation of microgrids (MGs) with high-quality power support, storage elements (e.g., batteries, flywheels, ultracapacitors, etc.) must be adequately accommodated as an integrated power system to ensure uninterrupted power supply during the maximum load demand [5–7].

An MG typically consists of a set of distributed micro-sources, energy storage systems (ESSs), and local loads, which are optimally planned for the profit of the customers [8, 9]. From the point of view of the utility, the MG can be implemented in two operations: (i) grid-connected (ii) isolated. In the grid-connected case, the MG is connected to the main utility grid, while the MG will be isolated from the utility when an interruption or disturbance has occurred in the utility. In an isolated MG, the variability of load demand and the fluctuations of RESs lead to frequency distraction; therefore, matching demand and generation is essential to maintain the reliability and stability of such integrated systems, being obtained by load frequency control (LFC). In the context of the secondary (or supplementary) LFC of MGs, various control methodologies were developed by contemporary researchers with promising results [10, 11].

Electric vehicles (EVs) are now gaining more attention, and it is predicted that the future of the automotive industry will be dominated by these devices [12–16]. The reason for their popularity is the fact that the EVs can be utilized as dispersed storage chargers that provide new facilities to the existing integrated power systems [17]. However, the high penetration of such technologies with a large magnitude of charging/discharging threatens the frequency stabilization of the MGs, and therefore, the EVs and RESs should be effectively coordinated to restore the frequency deviation within a tolerable range [18].

The integration of EVs and RESs brings new regulatory frequency issues to the MGs that should be addressed in the design of LFC. Up to now, many controllers such as H∞ control theory [19], MPC [20, 21], intelligent control [22], and non-integer [23, 24] were developed for the supplementary LFC problem of MGs. In [25], a model-free technique based on sliding theory was proposed for an islanded MG with EV devices. Moreover, in the context of plug-in hybrid EVs (PHEVs), an MPC scheme is designed in [26] to alleviate the frequency distraction of an integrated MG system. The control objectives pursued in [26] include the following: (***i***) to smooth the power output of wind and (***ii***) to decrease the number of needed PHEVs. The authors of [27] implemented a fuzzy control for frequency damping of hybrid ESSs/PVs/EVs. The results of [27] revealed that the suggested scheme efficiently handles the load and insolation variations.

According to the after-mentioned discussion, the integration of EVs and renewable resources imposes new uncertainties to MG reliability [28]. These frequency control challenges of the power system become more intensified when the charging EVs are implemented in the form of the multi-microgrids (MMGs) [29, 30]. Under such circumstances, the reliable operation of the power MGs can no longer be ensured using conventional deterministic control strategies. This necessitates the design of the advanced controllers for the realization of the MG operation requirements, such as fuzzy logic, predictive theory, and neural network [21, 31, 32]. In this chapter, a new SI-IT2-FPD/FPI controller is developed for supplementary LFC of an MMG. To further ameliorate the suggested controller's efficiency, an improved whale optimization algorithm (IWOA) is adopted for fine-tuning the PD/PI controller gains. In addition, different combinations of the two control coefficients (α_{pd} and α_{pi}) are investigated to show the performance of the SI-IT2-FLC under aggressive, moderate, and smooth cases. The suggested technique is applied to stabilize the frequency and tie-line power exchanges in the presence of load, wind, PV, and electric vehicles. The procedure design of the suggested controller is very simple as opposed to the model-based techniques [19, 33], and it can be implemented for a variety of practical applications. Lastly, real-time verifications based on OPAL-RT testbed are also made to demonstrate the capability of the designed SI-IT2-FPD/FPI controller in the practical field.

1.2 Study System and Modelling

1.2.1 The General Structure of Microgrid

Figure 1.1 shows a typical configuration of the MG, which consists of different distributed generation components like wind turbine (WT), photovoltaic (PV) cell, ESSs (e.g., battery ESS (BESS), flywheel ESS (FESS)), and local loads. In the integrated power plant, the control of the power grid is facilitated by the distribution management system, while the responsibility of the MG dispatch system is to satisfy the control requirements of the MG operation.

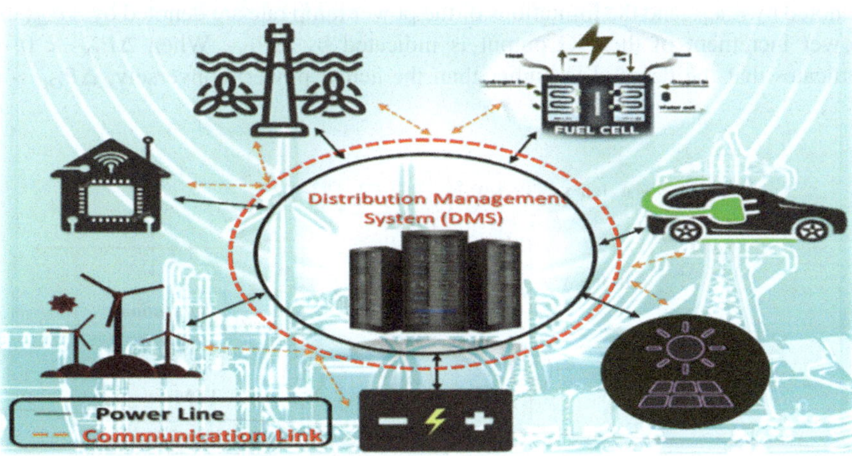

Fig. 1.1 General schematic of MGs

1.2.2 Model of the Electric Vehicle

The equivalent model of the EV considered in the simulation, proposed in [31, 32] as integrated into the form of MGs, is illustrated in Fig. 1.2. It is a popular model to investigate the specifications of an EV battery. In Fig. 1.2, T_e is the time constant, and Δu_e denotes the LFC action which is applied to the EV. Moreover, the inverter capacity is indicated by $\pm\mu_e$, and $\pm\delta_e$ denotes the limitations of the power ramp rate. The lower and upper of the EV's energy are shown by E_{min} and E_{max}, respectively. The charging/discharging power corresponding to the EV is represented by ΔP_E. When $\Delta P_E < 0$, the EV is in the charging mode, while $\Delta P_E > 0$ indicates the EV is discharging. In addition, for $\Delta P_E = 0$, the EV is in idle mode.

1.2.3 The Diesel Power System Model

The diesel generator (DG) is a small-scale generating system that is widely adopted in MG applications due to its valuable features like fast rapid speed, low maintenance, and long durability. The DG output can be regulated to respond to, instantly, the required power of the load disturbances. Also, the fluctuation of the RESs (i.e., PV and WT) can be compensated by the DG. The schematic diagram of the DG model is shown in Fig. 1.3, which illustrates the relationship between the LFC command and the power output of the DG. As displayed in this figure, the first-order inertia models are employed for the components of governor and generator.

In Fig. 1.3, Δf is the frequency deviation; Δu_{DG} denotes the LFC action signal; T_g and T_d represent the time constant coefficients corresponding to the governor and diesel, respectively; the state of the governor's valve is indicated by ΔX_G; R is the parameter which regulates the speed of DG; the limitations of power increment are denoted by $\pm\mu_{dg}$,and the limitations of the power ramp rate are denoted by $\pm\delta_{dg}$; the power increment of the DG output is indicated by ΔP_{DG}. When $\Delta P_{DG} < 0$, it indicates that the demand is higher than the actual power; conversely, $\Delta P_{DG} > 0$

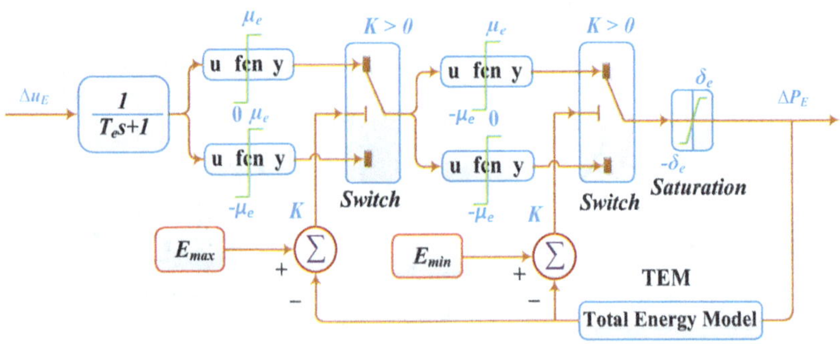

Fig. 1.2 The model of the EV in microgrid LFC

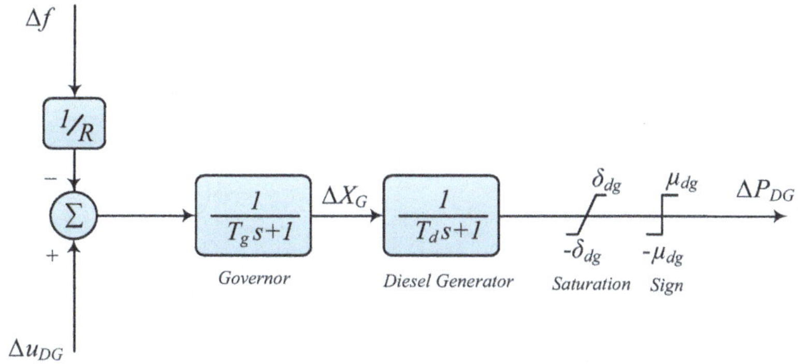

Fig. 1.3 The model of the diesel power system

indicates the generated power of the DG is more than the required power. Likewise, $\Delta P_{DG} = 0$ indicates that the demand and supply are in an equilibrium state [23, 32].

1.2.4 Model of Wind Turbine Generator (WTG) and Photovoltaic (PV)

Due to the time-variant and intermittent feature of the wind, the output of the wind turbine generator (WTG) changes based upon the wind speed and the physical characteristics of the turbine [30, 34]. Generally, the WTG can be considered as a disturbance in the MGs where its power fluctuations can be compensated by controlling the EVs and DG units.

A solar PV system produces power by converting solar energy, as an endless energy source, into electric energy. However, due to various reasons like the boundary and exterior contact and the small leakage current, the power loss will inevitably occur in such systems. The variability of the solar irradiance leads to the PV system operating in an intermittent condition; consequently, a random power source is often considered in the simulation applications to model the fluctuating feature of PV.

1.2.5 The Multi-microgrid System

For small-signal analysis, a multi microgrid (with two interconnected microgrids, i.e., MG-1 and MG-2) is modeled in a typical template as depicted in Fig. 1.4. As shown in Fig. 1.4, the MG-1 consists of DG, WT, PV, ESS, FESS, and EV, while only a DG unit is used in MG-2. The controller actions of each MG are designed

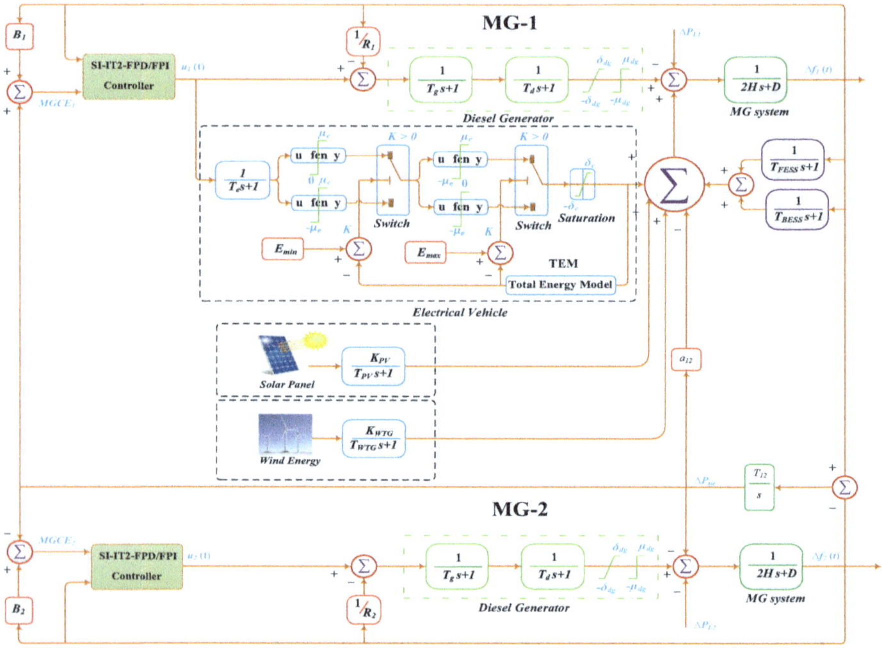

Fig. 1.4 Block diagram of the LFC model of multi-microgrid with EV

Table 1.1 The parameters of the MMG plant

Symbol and abbreviation	Values	Symbol and abbreviation	Values
T_g	2.000 s	K_{BESS}	−1/300
T_d	1.000 s	K_{PV}	1
T_e	1.000	K_{WTG}	1
R_1, R_2	3.000	T_{12}	0.2
B_1, B_2	0.9	a_{12}	-1
δ_{dg}, δ_e	0.025	Turbine radius, R	0.5 m
μ_{dg}, μ_e	0.012	Turbine height, h	2 m
E_{mini}	0.800	Maximum of power coefficient, C_p	0.195
E_{maxi}	0.950	Maximum of power coefficient, C_p	0.195
D	0.012	Optimum STR, λ_E	0.53
$2H$	0.200	Turbine inertia, J_τ	1.97 kg m^2
T_{FESS}	0.100 s	Water density, ρ_W	1000 kg/m^3
T_{BESS}	0.100 s	Water specific heat, C	4180 J/kg C
T_{PV}	1.8 s	Reynolds number constant, B	5
T_{WTG}	2 s	Impeller diameter, d	0628 m
K_{FESS}	−1/100	"s" means sec	

based on its MG control error (MGCE) and frequency deviation (Δf). The parameters adopted in the concerned MMG are listed in Table 1.1, as reported in [25].

1.3 The Suggested Structure of the SIT2-FLC

Figure 1.5 shows the graphical view of a new well-structured SIT2-FLC, namely, SI-IT2-FPD/FPI. The structure of Fig. 1.5 is constructed from the SI-IT2-FPD and SI-IT2-FPI controllers, and a detailed description of the scheme is presented in [35]. The input scaling factors (SFs) of IT2-FPD (k_{e1}) and IT2-FPI (k_{e2}) are used for normalizing the input to the universe of discourse. Here, these SFs are defined as $k_{e1} = 1/e_{max1}$ and $k_{e2} = 1/e_{max2}$, where e_{max1} and e_{max2} are the value of the maximum error for each of the SI-IT2-FPD and SI-IT2-FPI controllers, respectively. Therefore, the error (MGCE and Δf) will be converted, after normalization, into $\sigma_{o,\,pd}$ and $\sigma_{o,\,pi}$ which are the input of the SI-IT2-FPD and SI-IT2-FPI controllers, respectively. Then, the output of the SI-IT2-FPD ($\varphi_{o,\,pd}$) and output of the SI-IT2-FPI ($\varphi_{o,\,pi}$) are converted into control signal as:

$$u_{pd} = k_{u1}\left(k_{pd}\,\varphi_{o,pd} + k_d\,\frac{\varphi_{o,pd}}{dt}\right) \tag{1.1}$$

$$u_{pi} = k_{u2}\left(k_{pi}\,\varphi_{o,pi} + k_i\int \varphi_{o,pd}dt\right) \tag{1.2}$$

where k_{u1} and k_{u2} are the output SFs and are determined as $k_{u1} = k_{e1}^{-1}$ and $k_{u2} = k_{e2}^{-1}$. Moreover, $\{k_{pd},\, k_d\}$ and $\{k_{pi},\, k_i\}$ are the baseline PD and PI controller gains, respectively.

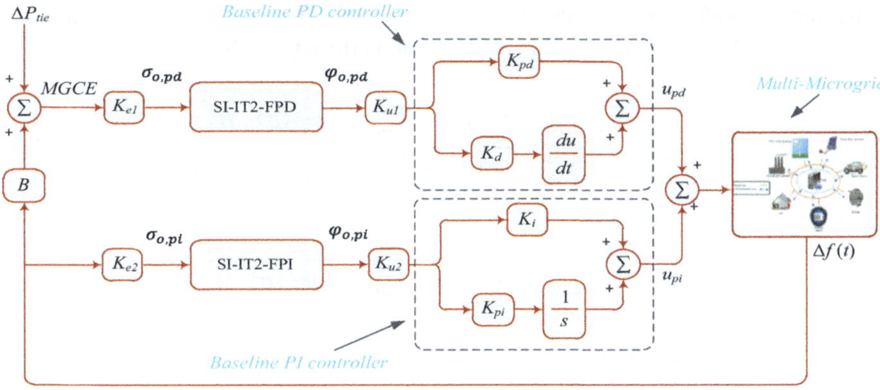

Fig. 1.5 Structure of the proposed SI-IT2-FPD/FPI

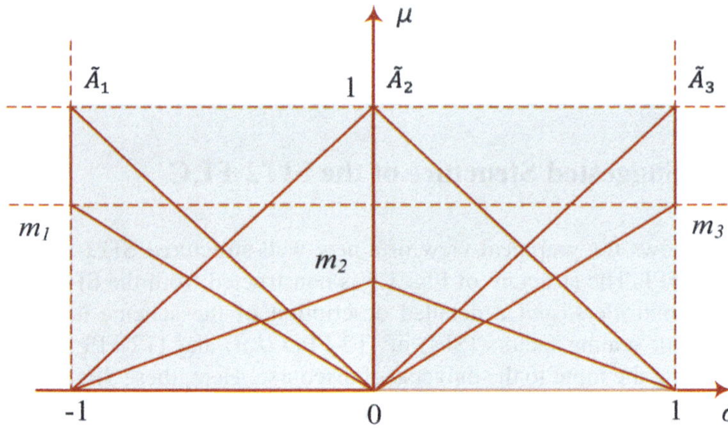

Fig. 1.6 Antecedent IT2-FSs of the IT2-FLC

1.3.1 Fundamental Concepts of Single-Input-IT2-FLC

The generic rule for each part of SI-IT2-FLC is defined as [36, 37]:

$$R_n : \text{if } \sigma \text{ is } \widetilde{A}_n.\text{Then } \varphi_o \text{ is } B_n \tag{1.3}$$

where the crisp consequents are represented by B_n with the values of $B_1 = -1$, $B_2 = 0$, and $B_3 = 1$. The membership functions (MFs) are formed by triangular IT2 fuzzy sets (IT2-FSs) $\widetilde{A}_n$, as shown in Fig. 1.6. The IT2-FSs can be defined in terms of lower MF ($\underline{\mu}_{\widetilde{A}_n}$) and upper MF ($\overline{\mu}_{\widetilde{A}_n}$). In Fig. 1.6, the lower MFs have the height of m_i's which creates extra freedom in the SI-IT2-FLC structure.

The symmetrical MFs are chosen, for simplicity, in this application with the following definition: $m_1 = m_3 = 1 - \alpha$ and $m_2 = \alpha$. Based on the center of type reduction methodology, the defuzzified crisp of an SI-IT2-FLC output is obtained as $\varphi_o = \left(\varphi_o^r + \varphi_o^l\right)/2$, where φ_o^r and φ_o^l are computed by [38, 39]

$$\varphi_o^l = \frac{\sum_{n=1}^{L} \overline{\mu}_{\widetilde{A}_n}(\sigma).\ B_n + \sum_{n=L+1}^{N} \underline{\mu}_{\widetilde{A}_n}(\sigma).\ B_n}{\sum_{n=1}^{L} \overline{\mu}_{\widetilde{A}_n}(\sigma) + \sum_{n=L+1}^{N} \underline{\mu}_{\widetilde{A}_n}(\sigma)} \tag{1.4}$$

$$\varphi_o^r = \frac{\sum_{n=1}^{R} \underline{\mu}_{\widetilde{A}_n}(\sigma).\ B_n + \sum_{n=R+1}^{N} \overline{\mu}_{\widetilde{A}_n}(\sigma).\ B_n}{\sum_{n=1}^{R} \underline{\mu}_{\widetilde{A}_n}(\sigma) + \sum_{n=R+1}^{N} \overline{\mu}_{\widetilde{A}_n}(\sigma)} \tag{1.5}$$

Here, the switching points (L, R) minimize and maximize the above equations.

Based on [40, 41], the fuzzy mapping of the SI-IT2-FLC $\varphi_o(\sigma)$ is determined as

$$\varphi_o(\sigma) = \sigma.k(|\sigma|) \tag{1.6}$$

where

$$k(\sigma) = \frac{1}{2}\left(\frac{1}{\alpha + \sigma - \alpha\sigma} + \frac{\alpha - 1}{\alpha\sigma - 1}\right) \tag{1.7}$$

1.3.2 Design of the SI-IT2-FLC

In this section, the design of the SI-IT2-FLCs with respect to the control curve (CC) is presented. Let $\varepsilon_o(\sigma) = \varphi_o(\sigma) - \sigma$, where $\varepsilon_o(\sigma)$ is the difference between unit mapping and IT2 fuzzy mapping. Then, different CCs can be obtained by:

(i) Aggressive CC (A-CC): when $0 < \alpha \leq \alpha_{c1}$, then $\varepsilon > 0$ for $\forall\sigma \in [0, 1)$ and $\alpha_{c1} = \left(3 - \sqrt{5}\right)/2$.
(ii) Smooth CC (S-CC): when $\alpha_{c2} < \alpha \leq 1$, then $\varepsilon < 0$ for $\forall\sigma \in [0, 1)$ and $\alpha_{c2} = \left(\sqrt{5} - 1\right)/2$.
(iii) Moderate CC (M-CC): when $\alpha_{c1} < \alpha \leq \alpha_{c2}$, then $\varepsilon < 0$ for $\forall\sigma \in [0.5, 1]$.

The curves of A-CC, S-CC, and M-CC are sketched in Fig. 1.7. The readers are referred to [35, 42] to find more description about the control curve features of the SI-IT2-FLC.

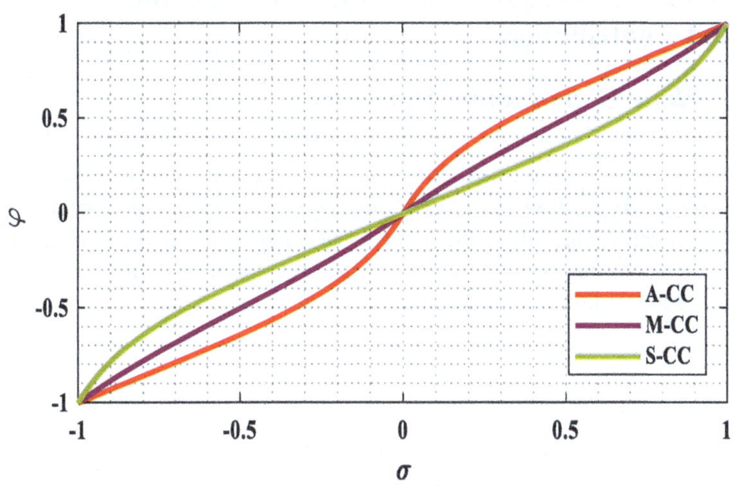

Fig. 1.7 Illustration of the S-CC, M-CC, and A-CC

For the suggested SI-IT2-FLC configuration (shown in Fig. 1.7), the following coefficient settings ($CS^{i}_{SI\text{-}IT2\text{-}FPD/FPI}$) are made for the i^{the} case.

(i) $CS^{1}_{SI\text{-}IT2\text{-}FPD/FPI}$ (α_{pd}= 0.25, α_{pi}= 0.25); both $\varphi_{o,\ PD}$ and $\varphi_{o,\ pi}$ are considered as A-CCs.

(ii) $CS^{2}_{SI\text{-}IT2\text{-}FPD/FPI}$ (α_{pd}= 0.25, α_{pi}= 0.5); $\varphi_{o,\ PD}$ and $\varphi_{o,\ pi}$ are considered with an A-CC and an M-CC, respectively.

(iii) $CS^{3}_{SI\text{-}IT2\text{-}FPD/FPI}$ (α_{pd}= 0.25, α_{pi}= 0.75); $\varphi_{o,\ PD}$ and $\varphi_{o,\ pi}$ are considered with an A-CC and an S-CC, respectively.

(iv) $CS^{4}_{SI\text{-}IT2\text{-}FPD/FPI}$ (α_{pd}= 0.5, α_{pi}= 0.25); both $\varphi_{o,\ PD}$ and $\varphi_{o,\ pi}$ are considered with an M-CC and an A-CC, respectively.

(v) $CS^{5}_{SI\text{-}IT2\text{-}FPD/FPI}$ (α_{pd}= 0.5, α_{pi}= 0.5); both $\varphi_{o,\ PD}$ and $\varphi_{o,\ pi}$ are considered as M-CCs.

(vi) $CS^{6}_{SI\text{-}IT2\text{-}FPD/FPI}$ (α_{pd}= 0.5, α_{pi}= 0.75); $\varphi_{o,\ PD}$ and $\varphi_{o,\ pi}$ are considered with an M-CC and an S-CC, respectively.

(vii) $CS^{7}_{SI\text{-}IT2\text{-}FPD/FPI}$ (α_{pd}= 0.75, α_{pi}= 0.25); $\varphi_{o,\ PD}$ and $\varphi_{o,\ pi}$ are considered with an S-CC and an A-CC, respectively.

(viii) $CS^{8}_{SI\text{-}IT2\text{-}FPD/FPI}$ (α_{pd}= 0.75, α_{pi}= 0.5); both $\varphi_{o,\ PD}$ and $\varphi_{o,\ pi}$ are considered with an S-CC and an M-CC, respectively.

(ix) $CS^{9}_{SI\text{-}IT2\text{-}FPD/FPI}$ (α_{pd}= 0.75, α_{pi}= 0.75); both $\varphi_{o,\ PD}$ and $\varphi_{o,\ pi}$ are considered as S-CCs.

Obviously, the output of the specific structured SI-IT2-FLC controller highly depends on the gains of the baseline PD/PI controllers. Thus, a new improved multi-objective heuristic technique is presented in the next section for the optimal setting of the gains to improve the performance of the considered SI-IT2-FLC controller.

1.4 Optimization Algorithm, Objective Function, and Contribution

1.4.1 Whale Optimization Algorithm

WOA is a new bio-inspired methodology (introduced by Mirjalili and Lewis [43]) that is formulated based on the unique foraging scheme of humpback whales. In the native bio-inspired algorithm, the improvisation process is accomplished in two phases, i.e., encircling with searching prey and spiral updating. The encircling behavior is modeled, based on the best solution, as below:

$$X_j^{t+1} = P_j^t - A.\left|B.P_j^t - X_j^t\right| \tag{1.8}$$

where $j = 1,2,...,D$, X_j^t is the position in the jth dimension at the tth iteration and P_j^t is the best solution in the jth dimension. The coefficient vectors of (1.8) (*A* and *B*) are determined by

$$A = 2.a.\ r - a \tag{1.9}$$

$$B = 2.\ r \tag{1.10}$$

where r is a random number in [0 1] and a is a control factor which is linearly reduced from 2 to 0 during the generations.

The movement characteristic of searching prey is figured out by

$$X_j^{t+1} = X_j^{rand} - A.\left|B.X_j^{rand} - X_j^t\right| \tag{1.11}$$

where X_j^{rand} is a random whale.

If |A| < 1, then the whale positions are guided in the path of the best solution by executing the encircling prey (1.8); otherwise, a random element is picked using searching prey (1.11) to enforce the exploration in the evaluation process.

The iterative process is characterized by a switching mechanism that executes the two phases (encircling and spiral) with a probability of 50%, simulated as

$$X_j^{t+1} = \begin{cases} P_j^t - A.\left|B.P_j^t - X_j^t\right| & \text{if } p < 0.5 \\ \left|P_j^t - X_j^t\right|.e^{bl}.\cos\ (2\pi l) + P_j^t & \text{if } p \geq 0.5 \end{cases} \tag{1.12}$$

where b is a constant factor that determines the spiral shape and p and l are random numbers in the ranges of [0 1] and [−1 1].

1.4.2 Improved Whale Optimization Algorithm [44]

Although the WOA has excellent capability in handling low-dimensional problems, this algorithm suffers from some inherent deficiencies and premature convergence in solving the multi/high-dimensional problems. To solve those weaknesses of the WOA, two perturbation mechanisms directed toward the best position are applied to the WOA individuals. In this approach, the polynomial mutation (PLM) [45] is improved by two maps: (*i*) the opposition-based learning (OBL) and (*ii*) the quasi OBL (QOBL) [46]. The PLM uses a polynomial probability distribution to change the current value of a variable to a neighboring value. Based on the PLM operator, the solutions are mutated from the best solution, described by

$$X_j^{t+1} = P_j^t + \bar{\delta}\Delta \tag{1.13}$$

$$\bar{\delta} = \begin{cases} (2\,r)^{(1/(q+1))} - 1 & \text{if } r < 0.5 \\ 1 - [2(1-r)]^{(1/(q+1))} & \text{Otherwise} \end{cases} \tag{1.14}$$

where q and Δ are the shape variable and the permissible perturbation, respectively. Under the mechanisms of OBL and QOBL, two distinct operations can be defined for obtaining the term Δ, given as

$$\Delta_{O,\ j} = l_j + u_j - X_j \tag{1.15}$$

$$\Delta_{QO,\ j} = rand\left(\frac{l_j + u_j}{2}, \Delta_{O,\ j}\right) \tag{1.16}$$

where l_j and u_j are the parameters which limit the jth decision variable.

For the convenience of readers, the pseudo-code of the IWOA approach is depicted in Algorithm 1.

1.4.3 Multi-objective IWOA

When an optimization problem includes several objectives that conflict with one another, the multi-objective scheme is preferred in such conditions such that the defined equality/inequality restrictions are satisfied [47, 48].

Definition 1.1 The generalized description of a multi-objective problem can be given as

$$Min : F(x) = \{\, f_1(x), f_2(x), \ldots, f_o(x)\}$$

$$Subject\ to : \begin{cases} g_i(x) \geq 0 & i = 1, 2, \ldots, K_{ueq} \\ h_i(x) = 0 & i = 1, 2, \ldots, K_{eq} \end{cases} \tag{1.17}$$

where o is the number of objectives functions and K_{ueq} and K_{eq} denote the inequality and equality restrictions, respectively. Likewise, g_i and h_i represent the ith inequality and equality constraints, respectively.

Definition 1.2 For an unconstrained optimization problem, an objective vector $X = (X_1, X_2, \ldots, X_D) \in \mathbb{R}^D$ dominates another objective vector $Y = (Y_1, Y_2, \ldots, Y_D) \in \mathbb{R}^D$ if the following two conditions are fulfilled:

Algorithm 1: The pseudo-code of the IWOA algorithm.

```
Initialize a set of search agents X_i (i=1, 2,..., n)
while (t < Max. Gen. ) do
    // Improvise a new search agent
    for j=1 to D do
        Update A, B, l, and p
        if 1 (p < 0.5)
            if 2 (|A| < +1)
                X_j^{t+1} = P_j^t - A . |B . P_j^t - X_j^t|
            else if 2 (|A| > +1)
                X_j^{t+1} = X_j^{rand} - A . |B . X_j^{rand} - X_j^t|
            end if2
        else if 1 (p ≥ 0.5)
            X_j^{t+1} = |P_j^t - X_j^t| . e^{bl} . cos (2πl) + P_j^t
        end if 1
    end for
    Update X* if there is a better solution t=t+1
    // Perform local search for the best whale
    if t < 0.25 * Max. Gen. then
        Generate a new solution by PLM operator based QOBL
        (see Eq. (13) and Eq. (16)) t=t+1
        If the profit of perturbed solution is better than X*, then
        replace the X* by the new position,
    end
    if t < 0.05 * Max. Gen. then
        Generate a new solution by PLM operator based OBL
        (see Eq. (13) and Eq. (15)) t=t+1
        If the profit of perturbed solution is better than X*, then
        replace the X* by the new position,
    end
    Record the best solution obtained so far
end
```

$$\begin{aligned} &\forall \quad j \in \{1, 2, \ldots, o\}, f_j(X) \leq f_j(Y) \\ &\exists \quad h \in \{1, 2, \ldots, o\}, \quad f_h(X) < f_h(Y) \end{aligned} \tag{1.18}$$

According to the above definition, the Pareto solutions can be characterized by non-dominated solutions in the decisive space. In this study, a multi-objective IWOA algorithm is used, employing the concept of Pareto optimization for the optimal setting of the gains embedded in the suggested controller. In this way, three objective functions are defined to design the gains of the established SIT2-FLC controller, given as

$$Min : J(x) = [J_1(x), J_2(x), J_3(x)]^T,$$

where

$$J_1 = \int_0^{t_{sim}} t\Delta f_1^2(t).dt \tag{1.19}$$

$$J_2 = \int_0^{t_{sim}} t\Delta f_2^2(t).dt \tag{1.20}$$

$$J_3 = \int_0^{t_{sim}} t\Delta P_{tie}^2(t).dt \tag{1.21}$$

where t_{sim} is the simulation time, Δf_1 and Δf_2 denote the frequency deviations in the MG-1 and MG-2, respectively, and ΔP_{tie} is the tie-line power.

To achieve a better solution among the optimal solutions, the fuzzy decision-making function with an MF is used to solve the multi-objective issue. The MF for each objective function is, mathematically, defined as

$$\mu_w^k = \begin{cases} 1 & f_i \le f_i^{min} \\ 0 & f_i \le f_i^{max} \\ \dfrac{f_i^{max} - f_i}{f_i^{max}(X) - f_i^{min}(X)} & f_i^{min} \le f_i \le f_i^{max} \end{cases} \tag{1.22}$$

In which f_i^{min} and f_i^{max} are the lowest and highest values of the ith objective function, respectively. In (1.22), $\mu_w = 0$ indicates the least satisfaction, while $\mu_w = 1$ indicates the maximum satisfaction.

For solving the optimization problem in a multi-objective manner, the fuzzy set mechanism is adopted as sketched in Fig. 1.8.

1.4.4 The Contribution of the Suggested Control Scheme

The goal of this work to develop a new model-free SI-IT2-FLC controller for the LFC problem of the MMG system with EVs. The following considerations are made by the suggested control framework that has a critical role in practical applications from a systematic perspective.

1. The new structured model-free controller is easy to be implemented in the HIL setup and can be adopted for other forms of the MG system.
2. Compared to the conventional fuzzy types (e.g., T1-FLC and IT2-FLC), the SI-IT2-FLC controller is simple to design as it does not need tuning the fuzzy membership functions and control rules of the FLCs.

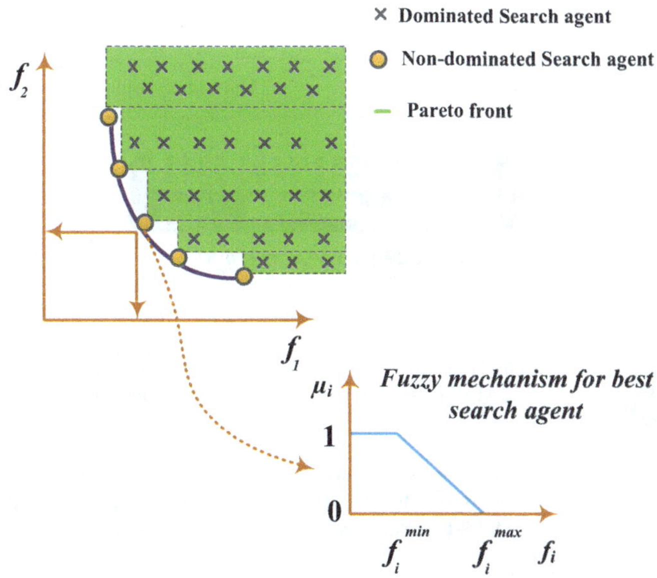

Fig. 1.8 The fuzzy set scheme employed for the Pareto set

3. The suggested control technique can be utilized for various configurations of the MG systems, with different loads, renewable sources, and grid typologies.
4. In addition to this, the suggested structured controller is designed and developed with low computational effort; this feature made its implementation feasible in real-word plants and online control purposes.
5. Finally, the comparative analysis under various MMG operations is made in a real-time platform to appraise the feasibility of the SI-IT2-FPD/FPI controller in the practical applications.

1.5 Simulation and Experimental Results

To ascertain the applicability and efficiency of the proposed controller, the multi-microgrid shown in Fig. 1.4 is simulated in MATLAB/Simulink software. Distinct supplementary LFC controllers are considered for each MG, and the performance of the established controllers is investigated against the variability of load demand and fluctuation of the RESs (wind, solar). The decisive controller gains are tuned by IWOA in a multi-objective approach, and then the effect of FOU parameters α_{pd} and α_{pi} of SIT2-FLC, for different cases of A-CC, M-CC, and S-CC, on control surface is investigated. To show the supremacy of the suggested controller, the case study is also examined with the T1-FPD/FPI and PD/PI controllers [49]. The lower and upper bounds for the optimal setting of the controller gains are set as 0 and 5, respectively, and the simulation time of $t_{sim} = 90$ s is considered with respect to

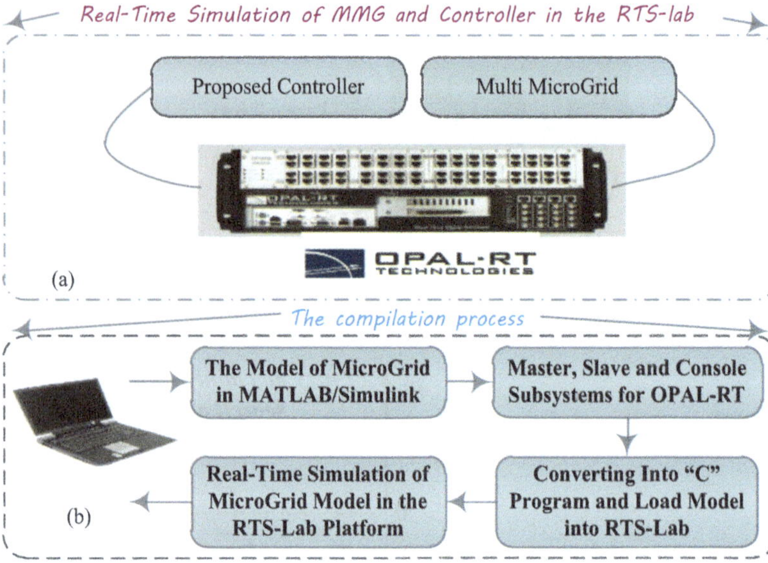

Fig. 1.9 The real-time experimental setup

the profiles of the RESs. In the current work, the following parameters are selected for application of the IWOA algorithm: the population size is 20, maximum generation is 50, p is a random number in [0 1], l is a random number in [−1 1], the variable for expressing the logarithmic spiral shape (b) is 1, and the shape parameter (q) is 20. The optimal parameters for the baseline PD/PI controllers, by employing multi-objective IWOA, are obtained as $k_{pd1} = 4.5781$, $k_{d1} = 4.8422$, $k_{pi1} = 2.4651$, and $k_{i1} = 5.0000$ for MG-1 and $k_{pd2} = 4.8550$, $k_{d2} = 4.9496$, $k_{pi2} = 1.2404$, and $k_{i2} = 4.8470$ for MG-2. To make a fair comparison, the optimal parameters are used for all of the controllers, and also the input SFs (k_{e1}, k_{e2}) and output SFs (k_{u1}, k_{u2}) are set to 1.

To appraise the feasibility of the suggested framework in the real-time context, a hardware-in-the-loop (HIL) analysis is conducted by employing the OPAL-RT testbed as sketched in Fig. 1.9. The HIL testbed is established to the impacts of the failures and delays that are not included in the off-line MATLAB environment. The main parts of the setup are as follows [50]:

1. An OPAL-RT platform (adopted as an RTS) which simulates the MMG sketched in Fig. 1.4.
2. A PC as the programming host in which the program code of MATLAB/Simulink will be run on the OPAL-RT.
3. A router to link whole the testbed components in a similar sub-network. In addition, the Ethernet ports are implemented to connect the OPAL-RT to DK60 board.

1.5.1 Scenario 1

In the first step, a constant load demand, i.e., $\Delta P_L = 0$, with the power fluctuation of WTG (ΔP_w) and PVG (ΔP_{pv}) is considered in the studied MMG. The real-data profile of ΔP_w [51] and ΔP_{pv} [52] is sketched in Fig. 1.10(a) and Fig. 1.10(b), respectively, which are taken from the data of an offshore wind farm in Sweden and the sun irradiance data of Aberdeen.

The comparative dynamic outputs (i.e., Δf_1, Δf_2, and ΔP) under the actions of SI-IT2-FPD/FPI (for cases 1, 2, 5, 6, 7, and 8 of the coefficient settings), T1-FPD/FPI, and PD/PI controllers are depicted in Fig. 1.11(a–c). It is perceived from the sub-figures of Fig. 1.11 that the peak value of responses in the structured SI-IT2 with $CS^{1,2}$ is relatively less than the other cases of the coefficient settings, T1-FPD/FPI, and PD/PI controllers. Moreover, smooth responses are obtained with the SI-IT2 with $CS^{5,6,8}$, T1-FPD/FPI, and PD/PI controllers. However, the SI-IT2 with CS^5 yielded more robust outcomes with less peak of overshoot than the others. On the other hand, the SI-IT2 with CS^7 experiences undesirable fluctuations with high

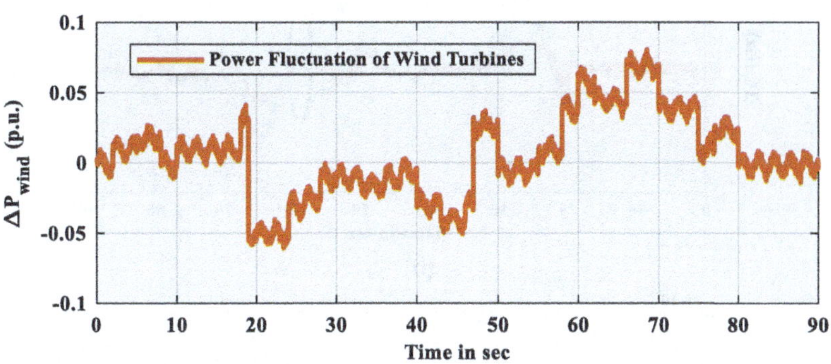

(a)

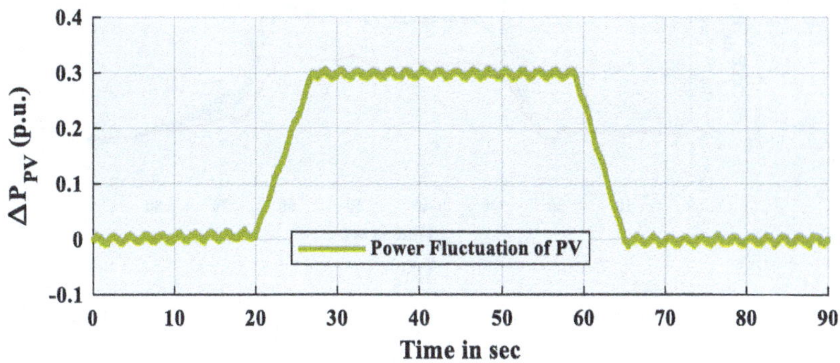

(b)

Fig. 1.10 Stochastic behavior of RESs (**a**): Profile of P_{WTG} (**b**): Profile of P_{PV}

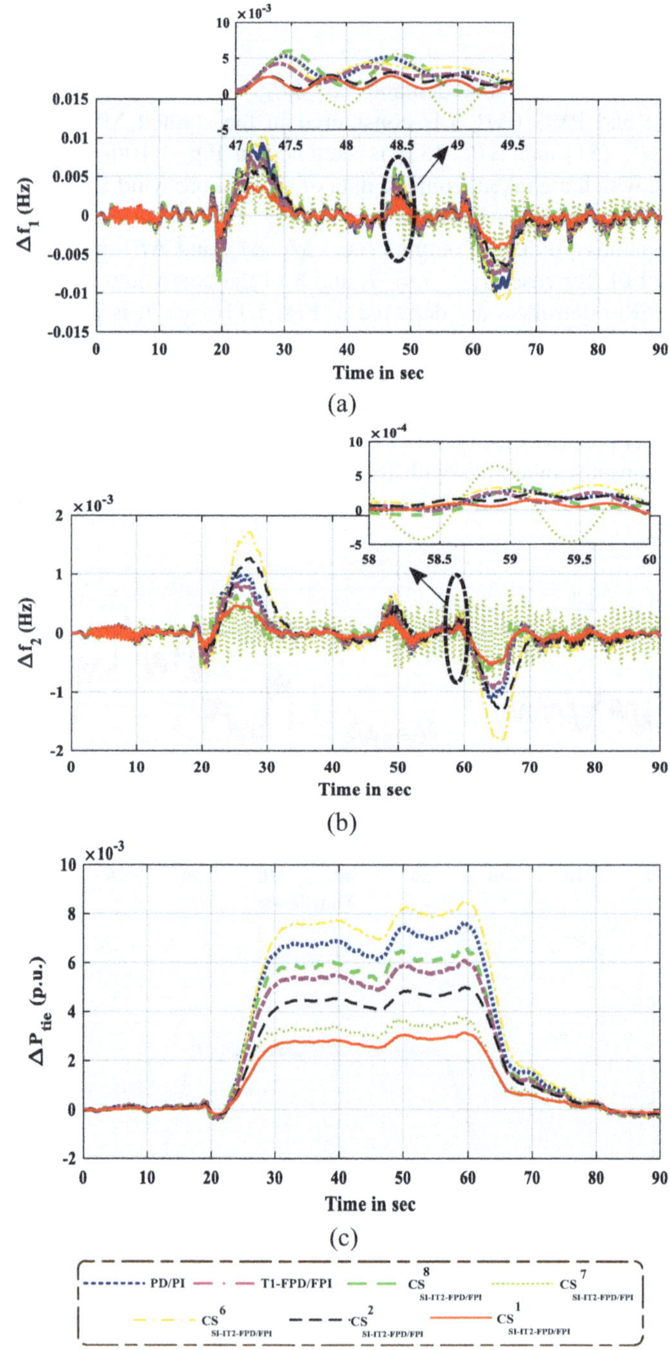

Fig. 1.11 Dynamic responses according to the power fluctuation of WTG and PV: (**a**) frequency deviation in MG-1, (**b**) frequency deviation in MG-2, (**c**) tie-line power deviation

Table 1.2 The value of the objective function for the different controllers under scenario 1

Controller	Coefficient settings	J
SI-IT2-FPD/FPI ($CS^{i}_{SI\text{-}IT2\text{-}FPD/FPI}$)	$CS^{1}_{SI\text{-}IT2\text{-}FPD/FPI}$	0.0201
	$CS^{2}_{SI\text{-}IT2\text{-}FPD/FPI}$	0.0522
	$CS^{3}_{SI\text{-}IT2\text{-}FPD/FPI}$	0.0945
	$CS^{4}_{SI\text{-}IT2\text{-}FPD/FPI}$	0.0258
	$CS^{5}_{SI\text{-}IT2\text{-}FPD/FPI}$	0.0740
	$CS^{6}_{SI\text{-}IT2\text{-}FPD/FPI}$	0.1497
	$CS^{7}_{SI\text{-}IT2\text{-}FPD/FPI}$	0.0390
	$CS^{8}_{SI\text{-}IT2\text{-}FPD/FPI}$	0.0892
	$CS^{9}_{SI\text{-}IT2\text{-}FPD/FPI}$	0.1906
T1-FPD/FPI	–	0.0609
PD/PI	–	0.1166

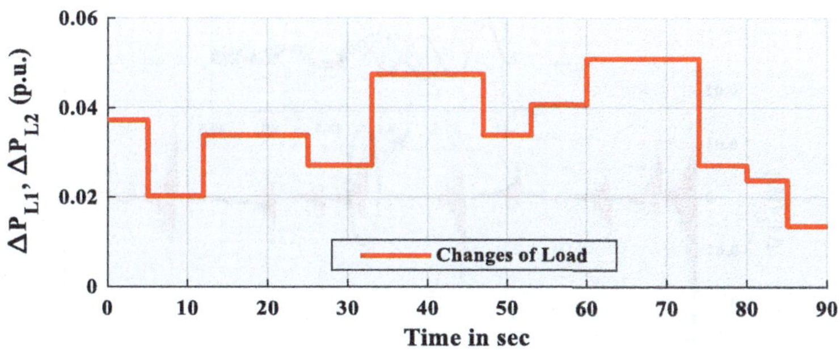

Fig. 1.12 Step variation of the load demand

magnitude throughout the simulation. It is noted from the real-time outcomes that the SI-IT2-FPD/FPI controller has aggressive action for the small values of α_{pd} and α_{pi}, while its action will be smoother when the values of α_{pd} and α_{pi} are increased. From the above discussions, it can be inferred that the coefficients α_{pd} and α_{pi} play a critical role in the performance of the SI-IT2-FPD/FPI controller; however, they do not function in the same way with gains of the baseline PD/PI controller. The values of the considered objective function for all the controllers are furnished in Table 1.2. As shown in this table, the lowest value of J is achieved with the SI-IT2-FPD/FPI for the cases of the structured SI-IT2 with CS^{1} than other controllers.

1.5.2 Scenario 2

In the current scenario, a multi-step load change, as typical disturbances, is imposed on each MG of the concerned system. The profile of the load changes is sketched in Fig. 1.12; the dynamic outputs with applications of the SI-IT2-FPD/FPI, T1-FPD/FPI, and PD/PI controllers are shown in Fig. 1.13(a–c).

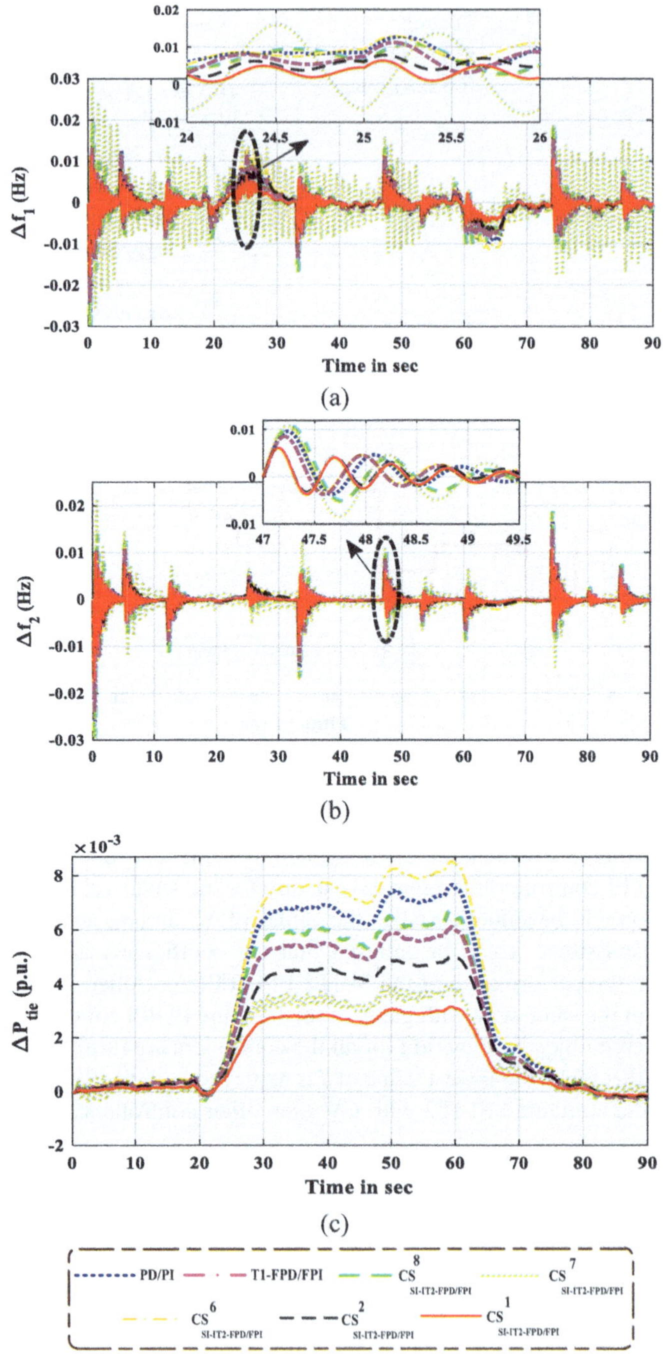

Fig. 1.13 Dynamic responses according to the power fluctuation of WTG and PV and load disturbances: (**a**) frequency deviation in MG-1, (**b**) frequency deviation in MG-2, (**c**) tie-line power deviation

Table 1.3 The value of the objective function for the different controllers under scenario 2

Controller	Coefficient settings	J
SI-IT2-FPD/FPI ($CS^{i}_{SI\text{-}IT2\text{-}FPD/FPI}$)	$CS^{1}_{SI\text{-}IT2\text{-}FPD/FPI}$	0.0344
	$CS^{2}_{SI\text{-}IT2\text{-}FPD/FPI}$	0.0657
	$CS^{3}_{SI\text{-}IT2\text{-}FPD/FPI}$	0.1082
	$CS^{4}_{SI\text{-}IT2\text{-}FPD/FPI}$	0.0625
	$CS^{5}_{SI\text{-}IT2\text{-}FPD/FPI}$	0.1017
	$CS^{6}_{SI\text{-}IT2\text{-}FPD/FPI}$	0.1771
	$CS^{7}_{SI\text{-}IT2\text{-}FPD/FPI}$	0.2398
	$CS^{8}_{SI\text{-}IT2\text{-}FPD/FPI}$	0.1415
	$CS^{9}_{SI\text{-}IT2\text{-}FPD/FPI}$	0.2376
T1-FPD/FPI	–	0.0817
PD/PI	–	0.1518

From the sub-figures of 13.1, it can be observed that the response fluctuations and the peak of the overshoot are relatively increased than the previous scenario (i.e., without consideration of load changes). In the current scenario, the transient behavior of the designed controllers is verified when a large load step is imposed on the system at $t = 60$ s. It is evident from Fig. 1.13 that by choosing a suitable CS, the SI-IT2-PD/PI controllers are still robust against the severe condition of change in load demand and fluctuation in RESs although a greater overshoot is observed. The values of the objective function for each case of the considered controllers are listed in Table 1.3. From this table, it is explicitly noted that the lowest value of the objective function is obtained by the CS^1 of the SI-IT2-PD/PI controller. However, the CS^7 and CS^9 of the structured SI-IT2 yielded poor performance with the biggest value of J among the controllers. On the other hand, the value of J is increased for all the controllers compared to when only solar and wind are applied. Thus, it can be concluded that the concerned load change degrades the performance of the MMG system.

1.5.3 Scenario 3

To assess the robustness of the designed LFC controllers, some critical parameters of the concerned MMG are changed as follows: $T_g = +30\%$; $T_e = -25\%$; $B = +25\%$; $R = -35\%$; $D = +20\%$; $H = -30\%$; $T_{FESS} = +25\%$; $T_{BESS} = +35\%$. The curves of the frequency fluctuation for the specific scenario are given in Fig. 1.14.

It can be observed from the subfigures of Fig. 1.14 that the response frequency of the MMG under the extreme parameter variation is much more fluctuating than when the parameters are at the operation condition. In addition, the peaks of overshoot for the different controllers are relatively increased under the concerned scenario.

Remark 1.1 It is worth mentioning that the suggested SI-IT2-FLC controller has a more straightforward design and less execution time in comparison with its type-1 and conventional counterparts. Moreover, it is notable that though the SI-IT2-FLC

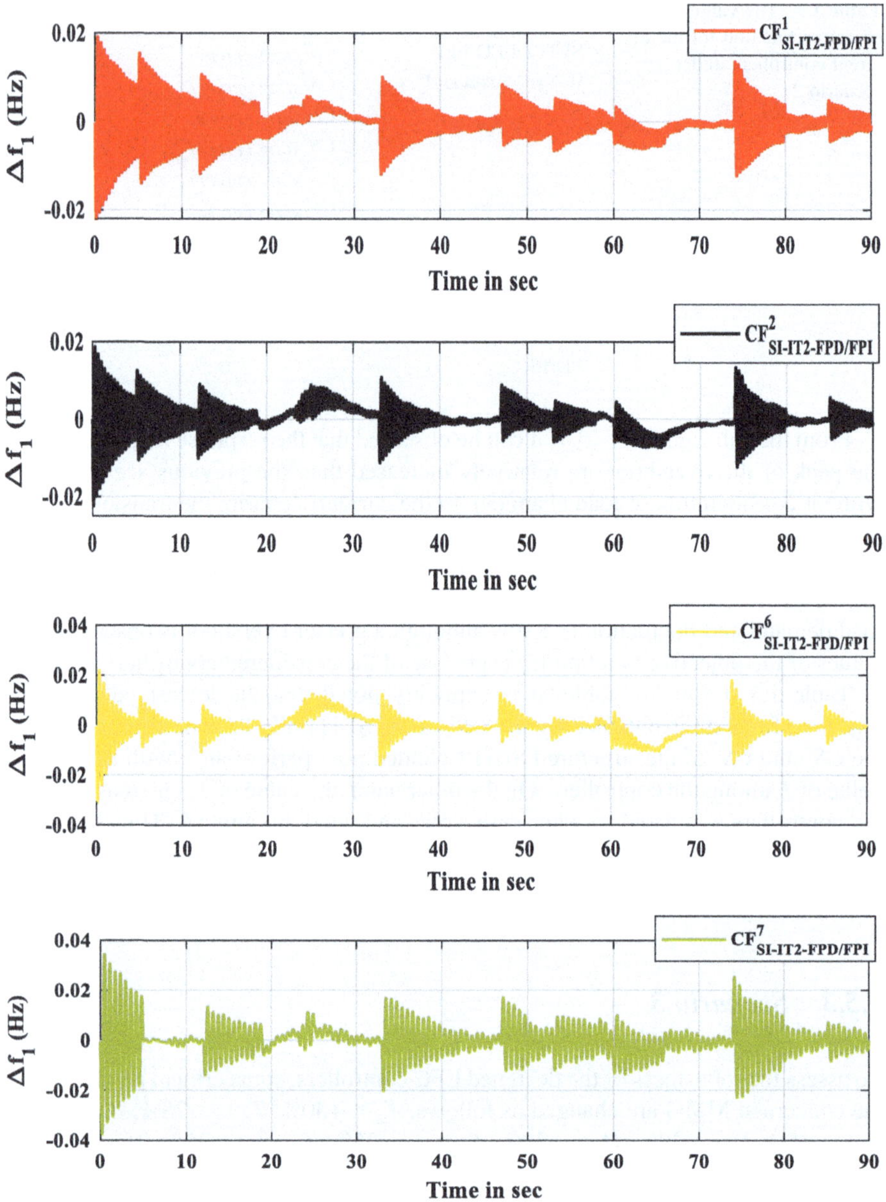

Fig. 1.14 Frequency deviation in MG-1 according to scenario 3

controller has not shown significant improvement in the dynamic responses than the T1-FLC and PID controllers (which were discussed in the three scenario studies), its obvious advantage is reducing the computational burden, and straightforward

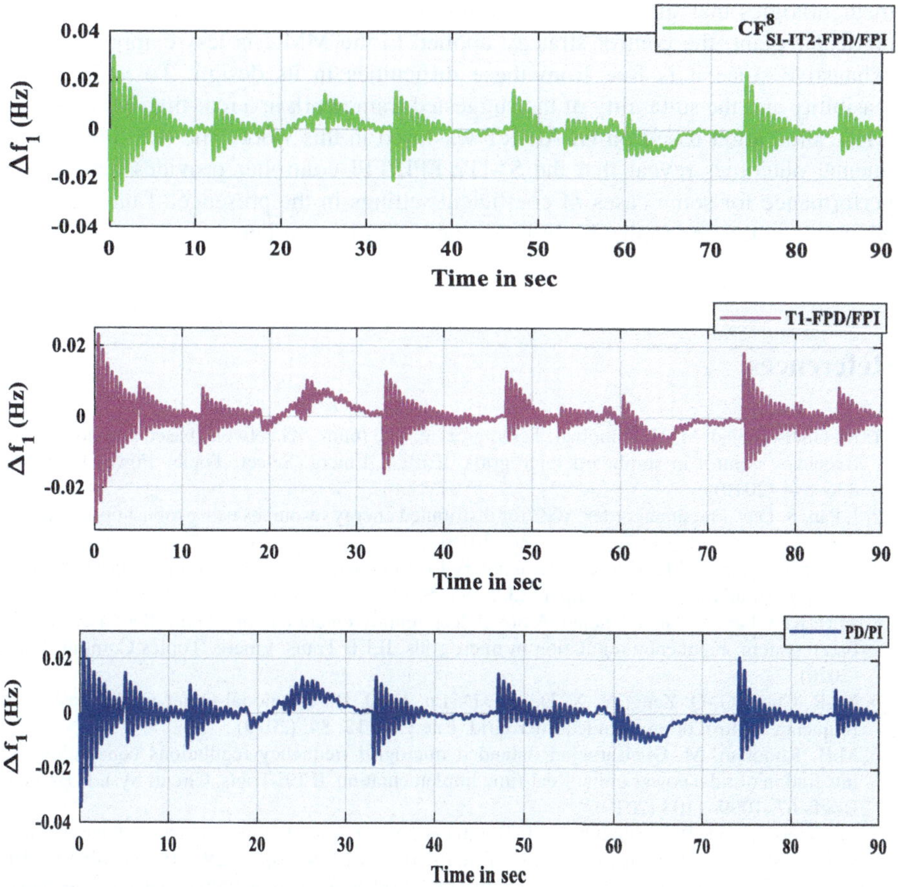

Fig. 1.14 (continued)

adjusting provides a practical and applicable alternative for hardware implementations of the aforesaid technical problems.

1.6 Conclusions

In this chapter, a new structured SI-IT2-FPD/FPI controller was developed for the LFC issue of an MMG with the application of wind generation, PV, and EV. To ameliorate the performance of the suggested SI-IT2-FLC controller against the said uncertainties, the secondary LFC of MMG was designed in two phases. Initially, a multi-objective IWOA is proposed for the optimal tuning of the parameters embedded in the baseline PD/PI controller. Then, by tuning a single coefficient that shapes the FOU, various CSs have been generated. In contrast to the model-based methodologies that suffer from the complexity of the mathematical modeling of a particular plant, the control strategy applied to the MMG is less computationally exhaustive since it is free from these difficulties in its design. To validate the feasibility and the suitability of the suggested framework in a real-time MG testbed, a HIL simulation based on OPAL-RT was used in this work. The real-time experimental outcomes reveal that the SI-IT2-FPD/FPI controller provides better LFC performance for some cases of coefficient settings in the presence of uncertainties when it is compared to the T1-FPD/FPI and PD/PI controllers.

References

1. M. Gheisarnejad, M.-H. Khooban, T. Dragičević, The future 5G network-based secondary load frequency control in shipboard microgrids. IEEE J. Emerg. Select. Topics Power Electr. **8**, 836–844 (2019)
2. I. Pan, S. Das, Fractional order AGC for distributed energy resources using robust optimization. IEEE Trans. Smart Grid **7**, 2175–2186 (2015)
3. M. Gheisarnejad, M.H. Khooban, Secondary load frequency control for multi-microgrids: HiL real-time simulation. Soft. Comput. **23**, 5785–5798 (2019)
4. M.H. Khooban, M. Gheisarnejad, A novel deep reinforcement learning controller based type-II fuzzy system: Frequency regulation in microgrids. IEEE Trans. Emerg. Topics Comput. Intel. (2020)
5. M.-R. Chen, G.-Q. Zeng, Y.-X. Dai, K.-D. Lu, D.-Q. Bi, Fractional-Order model predictive frequency control of an islanded microgrid. Energies, **12**, 84, (2019)
6. M.H. Khooban, M. Gheisarnejad, Islanded microgrid frequency regulations concerning the integration of tidal power units: Real-time implementation. IEEE Trans. Circuit Syst. II Express Briefs **67**, 1099–1103 (2019)
7. M. Gheisarnejad, P. Karimaghaee, J. Boudjadar, M.-H. Khooban, Real-time cellular wireless sensor testbed for frequency regulation in smart grids. IEEE Sensors J. **19**, 11656–11665 (2019)
8. I. Pan, S. Das, Fractional order fuzzy control of hybrid power system with renewable generation using chaotic PSO. ISA Trans. **62**, 19–29 (2016)
9. K.S. Rajesh, S.S. Dash, Load frequency control of autonomous power system using adaptive fuzzy based PID controller optimized on improved sine cosine algorithm. J. Ambient. Intell. Humaniz. Comput. **10**, 2361–2373 (2019)
10. M.H. Fini, M.E.H. Golshan, Determining optimal virtual inertia and frequency control parameters to preserve the frequency stability in islanded microgrids with high penetration of renewables. Electr. Power Syst. Res. **154**, 13–22 (2018)
11. H. Wang, G. Zeng, Y. Dai, D. Bi, J. Sun, X. Xie, Design of a fractional order frequency PID controller for an islanded microgrid: A multi-objective extremal optimization method. Energies **10**, 1502 (2017)
12. M.H. Khooban, T. Niknam, M. Sha-Sadeghi, Speed control of electrical vehicles: A time-varying proportional–integral controller-based type-2 fuzzy logic. IET Sci. Measur. Technol. **10**, 185–192 (2016)
13. M. Rahmani-Andebili, M. Bonamente, J.A. Miller, Charging management of plug-in electric vehicles in San Francisco applying Monte Carlo Markov chain and stochastic model predictive control and considering renewables and drag force. IET Gener. Trans. Distrib. (2020)
14. M. Rahmani-Andebili, M. Fotuhi-Firuzabad, An adaptive approach for PEVs charging management and reconfiguration of electrical distribution system penetrated by renewables. IEEE Trans. Ind. Inf. **14**, 2001–2010 (2017)

15. M. Rahmani-Andebili, Studying the effects of plug-in electric vehicles on the real power markets demand considering the technical and social aspects, in *Planning and Operation of Plug-In Electric Vehicles* (Springer, 2019), pp. 1–21
16. M.-H. Khooban, M. Gheisarnejad, N. Vafamand, J. Boudjadar, Electric vehicle power propulsion system control based on time-varying fractional calculus: Implementation and experimental results. IEEE Trans. Intel. Vehic. **4**, 255–264 (2019)
17. M. Rahmani-Andebili, Spinning reserve capacity provision by the optimal fleet management of plug-in electric vehicles considering the technical and social aspects, in *Planning and Operation of Plug-In Electric Vehicles* (Springer, 2019), pp. 49–74
18. S. Falahati, S. A. Taher, M. Shahidehpour, Grid secondary frequency control by optimized fuzzy control of electric vehicles. IEEE Trans. Smart Grid (2017)
19. T. Goya, E. Omine, Y. Kinjyo, T. Senjyu, A. Yona, N. Urasaki, et al., Frequency control in isolated island by using parallel operated battery systems applying H∞ control theory based on droop characteristics. IET Renew. Power Gener. **5**, 160–166 (2011)
20. J. Yang, Z. Zeng, Y. Tang, J. Yan, H. He, Y. Wu, Load frequency control in isolated microgrids with electrical vehicles based on multivariable generalized predictive theory. Energies **8**, 2145–2164 (2015)
21. S. Kayalvizhi, D.M.V. Kumar, Load frequency control of an isolated micro grid using fuzzy adaptive model predictive control. IEEE Access **5**, 16241–16251 (2017)
22. H. Bevrani, F. Habibi, P. Babahajyani, M. Watanabe, Y. Mitani, Intelligent frequency control in an AC microgrid: Online PSO-based fuzzy tuning approach. IEEE Trans. Smart Grid **3**, 1935–1944 (2012)
23. M.-H. Khooban, T. Niknam, M. Shasadeghi, T. Dragicevic, F. Blaabjerg, Load frequency control in microgrids based on a stochastic noninteger controller. IEEE Trans. Sustain. Energy **9**, 853–861 (2018)
24. K. Nosrati, H.R. Mansouri, H. Saboori, Fractional-order PID controller design of frequency deviation in a hybrid renewable energy generation and storage system. CIRED-Open Access Proc. J. **2017**, 1148–1152 (2017)
25. M.-H. Khooban, Secondary load frequency control of time-delay stand-alone microgrids with electric vehicles. IEEE Trans. Ind. Electron. **65**, 7416–7422 (2018)
26. J. Pahasa, I. Ngamroo, Coordinated control of wind turbine blade pitch angle and PHEVs using MPCs for load frequency control of microgrid. IEEE Syst. J. **10**, 97–105 (2016)
27. M. Datta, T. Senjyu, Fuzzy control of distributed PV inverters/energy storage systems/electric vehicles for frequency regulation in a large power system. IEEE Trans. Smart Grid **4**, 479–488 (2013)
28. M. Rahmani-Andebili, Studying the effects of optimal fleet management of plug-in electric vehicles on the unit commitment problem considering the technical and social aspects, in *Planning and Operation of Plug-In Electric Vehicles* (Springer, 2019), pp. 23–47
29. A.A. El-Fergany, M.A. El-Hameed, Efficient frequency controllers for autonomous two-area hybrid microgrid system using social-spider optimiser. IET Gener. Transm. Distrib. **11**, 637–648 (2017)
30. S.K. Pandey, S.R. Mohanty, N. Kishor, J.P.S. Catalão, Frequency regulation in hybrid power systems using particle swarm optimization and linear matrix inequalities based robust controller design. Int. J. Electr. Power Energy Syst. **63**, 887–900 (2014)
31. M.-H. Khooban, T. Niknam, F. Blaabjerg, P. Davari, T. Dragicevic, A robust adaptive load frequency control for micro-grids. ISA Trans. **65**, 220–229 (2016)
32. M.H. Khooban, T. Niknam, F. Blaabjerg, T. Dragičević, A new load frequency control strategy for micro-grids with considering electrical vehicles. Electr. Power Syst. Res. **143**, 585–598 (2017)
33. H. Bevrani, M.R. Feizi, S. Ataee, Robust frequency control in an islanded microgrid: Hinf and μ-synthesis approaches. IEEE Trans. Smart Grid **7**, 706–717 (2016)

34. K.S. Rajesh, S.S. Dash, Load frequency control of autonomous power system using adaptive fuzzy based PID controller optimized on improved sine cosine algorithm. J. Ambient. Intell. Humaniz. Comput., 1–13 (2018)
35. A. Sarabakha, C. Fu, E. Kayacan, T. Kumbasar, Type-2 fuzzy logic controllers made even simpler: From design to deployment for UAVs. IEEE Trans. Ind. Electron. **65**, 5069–5077 (2018)
36. R. Heydari, M. Gheisarnejad, M.H. Khooban, T. Dragicevic, F. Blaabjerg, Robust and fast voltage-source-converter (VSC) control for naval shipboard microgrids. IEEE Trans. Power Electron. **34**, 8299–8303 (2019)
37. M. Gheisarnejad, J. Boudjadar, M.-H. Khooban, A new adaptive type-II fuzzy-based deep reinforcement learning control: Fuel cell air-feed sensors control. IEEE Sensors J. **19**, 9081–9089 (2019)
38. M. Gheisarnejad, H. Mohammadi-Moghadam, J. Boudjadar, M.H. Khooban, Active power sharing and frequency recovery control in an islanded microgrid with nonlinear load and nondispatchable DG. IEEE Syst. J. **14**, 1058–1068 (2019)
39. A. Sarabakha, C. Fu, E. Kayacan, T. Kumbasar, Type-2 fuzzy logic controllers made even simpler: From design to deployment for UAVs. IEEE Trans. Ind. Electron. **65**, 5069–5077 (2017)
40. T. Kumbasar, Robust stability analysis and systematic design of single-input interval type-2 fuzzy logic controllers. IEEE Trans. Fuzzy Syst. **24**, 675–694 (2016)
41. M. Mehndiratta, E. Kayacan, T. Kumbasar, Design and experimental validation of single input type-2 fuzzy PID controllers as applied to 3 DOF helicopter testbed, in *2016 IEEE International Conference on Fuzzy Systems (FUZZ-IEEE)*, 2016, pp. 1584–1591
42. T. Kumbasar, H. Hagras, A gradient descent based online tuning mechanism for PI type single input interval type-2 fuzzy logic controllers, 1–6
43. S. Mirjalili, A. Lewis, The whale optimization algorithm. Adv. Eng. Softw. **95**, 51–67 (2016)
44. M. Gheisarnejad, P. Karimaghaee, J. Boudjadar, M.-H. Khooban, Real-time cellular wireless sensor testbed for frequency regulation in smart grids. IEEE Sensors J. (2019)
45. G.-Q. Zeng, J. Chen, L.-M. Li, M.-R. Chen, L. Wu, Y.-X. Dai, et al., An improved multi-objective population-based extremal optimization algorithm with polynomial mutation. Inf. Sci. **330**, 49–73 (2016)
46. Z. Seif, M.B. Ahmadi, An opposition-based algorithm for function optimization. Eng. Appl. Artif. Intell. **37**, 293–306 (2015)
47. I. Pan, S. Das, Kriging based surrogate modeling for fractional order control of microgrids. IEEE Trans. Smart Grid **6**, 36–44 (2015)
48. M.-H. Khooban, T. Dragicevic, F. Blaabjerg, M. Delimar, Shipboard microgrids: A novel approach to load frequency control. IEEE Trans. Sustain. Energy **9**, 843–852 (2018)
49. V. Kumar, K.P.S. Rana, P. Mishra, Robust speed control of hybrid electric vehicle using fractional order fuzzy PD and PI controllers in cascade control loop. J. Franklin Inst. **353**, 1713–1741 (2016)
50. H. Zhang, Y. Zhang, C. Yin, Hardware-in-the-loop simulation of robust mode transition control for a series–parallel hybrid electric vehicle. IEEE Trans. Veh. Technol. **65**, 1059–1069 (2016)
51. www.winddata.com. [Online; accessed 10.10.14]
52. www.solargis.info/doc/solar-and-pv-data. [Online;accessed 10.10.14]

Chapter 2
Distributed Noise-Resilient Control for DC Microgrids Under Dynamic Communication Topology

Jingang Lai and Xiaoqing Lu

Abstract This chapter proposes a discrete-time distributed mean-square consensus cooperation scheme that can achieve DC bus voltage restoration and maintain proportional current sharing of DC microgrids in mean square via a sparse communication network subject to dynamic communication topology and multiplicative noise disturbances. Since the cyber networks are exposed to multiplicative noise disturbance and the switching of dynamic communication topologies, it terribly reduces the stability and quality of whole system. To eliminate the adverse effects of dynamic communication topology and Brownian noise disturbances, a robust state-dependent multiplicative noise resiliency distributed resilient control algorithm is developed for DC microgrids. Through adopting stochastic stability theory and Lyapunov function, the sufficient conditions considering dynamic communication topology and noise interferences are derived to guarantee the stability operation of the whole closed-loop system. As a result, the suggested method decreases the sensitivity of the system to failures and increases its reliability. Finally, a DC microgrid test system in MATLAB/Simulink is utilized to verify the effectiveness of the proposed controller design scheme.

Keywords Distributed discrete-time control · Mean-square consensus · Dynamic communication topology · Multiplicative noise · Current sharing · Voltage restoration · DC microgrid

Nomenclature

N	Number of DGs in a DC microgrid
v_i^{nom}	The output voltage nominal set point of DG_i

J. Lai (✉)
E.ON Energy Research Center, RWTH Aachen University, Aachen, Germany

X. Lu
School of Electrical Engineering and Automation, Wuhan University, Wuhan, China
e-mail: luxq@whu.edu.cn

M. Rahmani-Andebili (ed.), *Design, Control, and Operation of Microgrids in Smart Grids*, Power Systems, https://doi.org/10.1007/978-3-030-64631-8_2

v_i^{ref}	The desired value voltage of DG_i
v_i^*	The local control voltage of DG_i
i_i^{ref}	The current reference value of DG_i
i_i	The measured current of DG_i
R_i^{v}	The virtual resistance of DG_i
Δv_i	The maximum allowed voltage deviation
$i_{i\max}$	The maximum allowed current
v_i^{C}	The capacitance voltage of LC filter
i_i^L	The inductor current of LC filter
R_i^{f}	The filter resistance
L_i^{f}	The filter inductor
C_i^{f}	The filter capacitor
d_i^{buck}	The duty cycle of the buck converter i
$\mathcal{A}_{\mathcal{G}}$	The adjacency matrix of $\mathcal{G}$
$\mathcal{L}_{\mathcal{G}}$	The Laplacian matrix of $\mathcal{G}$
$\mathcal{D}_{\mathcal{G}}$	The degree matrix of $\mathcal{G}$
$\mathscr{B}$	The degree matrix of $\overline{\mathcal{G}}$
λ_i	The ith eigenvalue of the matrix $\mathcal{L} + \mathscr{B}$

2.1 Introduction

Nowadays, with the various sources of distributed generation (DG) seamlessly integrated into the modern power system, rapidly developed microgrids provide a promising application on localized DG utilization, which increases energy efficiency with no reliance on the grid and optimal total cost of ownership [1, 2, 28]. Consequently, microgrids play an important role in increasing the resilience of critical power infrastructure [3], which needs a modern control and management system to improve the effectiveness of their operation and provide global stability [4]. Microgrids, as subsystems in smart grids, can be operated in both grid-connected and islanded modes [5, 29]. A microgrid is a cyber-physical system, which has electrical networks integrated with DGs and energy storage systems (ESSs) and communication networks for coordinated control and operation of DGs [6]. In the electrical systems, DGs and ESSs are operated to deal with the power variations from renewable energy sources and maintain power supply for local loads [7]. In the communication systems, distributed cooperative control based on multi-agent systems is adopted among controllable units to increase the flexibility, reliability, and scalability of microgrid systems [8]. More recently, DC microgrids have gained much attention due to their increased efficiency in delivering power and flexibility for the integration of distributed generators with DC nature (e.g., photovoltaic and battery energy storage systems) [9, 10]. As one of the major control objectives in DC microgrids, proper voltage regulation while satisfying the proportional power sharing among DGs is of paramount value.

For microgrids in islanded conditions, droop control is widely adopted to realize the abovementioned two control objectives in a decentralized manner. The basis of the droop-based methods for the DC microgrids is in the linear decrement of the voltage reference value by increasing the output current. In DC microgrids, the output current or power can be used as the feedback signal [7]. However, droop control methods have several drawbacks. For example, droop-based methods are not robust against load changes and nonlinear or unbalanced loads, which can result in instability. These methods also create a compromise between the accuracy of current distribution and voltage regulation. Increasing the droop coefficients increases the accuracy of current distribution and also increases the damping of the system, but on the other hand, it causes voltage deviation, which again raises the need to use a secondary controller. Slow transient response, poor performance against resistive-inductive lines, voltage steady-state error, and inability to attain a coordinated performance of multiple components with different characteristics are the main drawbacks of droop-based control methods [9–12]. Even though droop control is communication-free and easy to implement, the simple control technique is inherently incapable of achieving accurate load sharing and voltage regulation simultaneously. To eliminate the control deviations of droop control, secondary controls need to be implemented.

According to the control structure, secondary control strategies are generally categorized into three types, i.e., centralized, decentralized, and distributed matters [11, 12]. Due to the various advantages of distributed control technology (e.g., the low cost of communication with low-bandwidth communication links, robustness to controller single point of failure, feasible for the plug-and- play capability) [13–15], the distributed control approach, which only needs all DGs communicate and share information with their neighbors through sparse communication networks, has attracted a great deal of attention. Recently, different aspects of distributed secondary control of microgrid such as event-triggered communication [16, 17], state-independent noise resiliency [18], and finite-time consensus [19] have been addressed in the literature.

As a matter of fact, the communication network of a distribution microgrid is generally subject to inherent communication noise disturbances, which may endanger the stability of microgrids and degrade their dynamic performance. Therefore, it is necessary to build a distributed cooperative control scheme for DC microgrid systems by considering communication noise disturbances. Most recently, some interesting results on the effects of communication noise disturbances for AC microgrid stability have been investigated in [16–20]. For instance, a noise-resiliency distributed controller is designed for a radial AC microgrid in [13]. Moreover, communication topologies of microgrid systems can be dynamic switching from one pattern to another due to different communication requirements, grid structure changes, or environmental disturbances. For example, the dynamic switching of communication topologies will occur when a communication link between two DGs is either provisionally or eternally broken by external disturbances. Furthermore, due to intermittent feature of distributed renewable energy resources, removal or addition of DGs in a microgrid (referred to as plug-and-play functionality) also leads to the switching of communication topologies [20]. Thus, it is essential to develop distributed controllers which can be suitable for switching communication topologies. Although some simulation cases with switching dynamic communication topologies caused by some link failures or unit plug-in and plug-out are performed in [21], they do not reveal theoretically how switching dynamic topologies affect stability of microgrids. In order to simplify the analysis, the works in [22, 23] have considered switching communication topologies, but it ignores the influence of noise interference on the stability of the system. Based on the above analysis, we can see clearly that in most of the exiting results, communication noise disturbances and switching topologies are not simultaneously considered to achieve distributed cooperative control of DC microgrids. Furthermore, the switching-dynamic-topology and communication noises are coupled with each other in real applications, which has increased so that these issues will become more crucial and challenging.

Taking into account all aforementioned issues, in this chapter, a discrete-time distributed mean-square consensus control scheme considering multiplicative noises will be proposed to properly work irrespective of time-varying, restricted, and unreliable communication networks while preserving the voltage restoration and autonomous current sharing. Different from previous works, the main salient features of the proposed designs are as follows:

1. Since the communication rate is limited in real applications, the communication equipment and network are actually a discrete-time system. Thus, this chapter proposes a discrete-time distributed mean-square consensus control scheme to achieve DC bus voltage restoration and maintain proportional current sharing among massive DGs in mean square, even if the sparse communication network suffers from noise disturbances and switching topology. It makes our result different from existing ideal and continuous-time communication over microgrids [12–14]. Furthermore, the network multiplicative noise disturbances and switching topology are simultaneously considered in the proposed stochastic discrete-time control system, which makes our results more meaningful yet practical.
2. Compared with the existing state-independent additive noise-resilient distributed control algorithms for AC microgrid in [7, 16, 18, 27], a novel state-dependent multiplicative noise-resilient distributed control algorithm is developed; meanwhile, it is also robustness against microgrid topological uncertainties and maintaining the plug-and-play functionality of DGs. Furthermore, by using the stochastic analysis techniques combined with the Lyapunov function, sufficient conditions on the control gain designed considering noise disturbances and switching dynamic communication topology are established to guarantee the stability and reliability of the whole systems. Since stability issues may appear if discrete-time feature of the communication part is considered, we differentiate

the time step of communication part and controller part so as to emulate a more realistic system.

3. Because of plug-and-play capability of microgrids, microgrid's physical and communication structures can be time-varying. For this reason, in this chapter, a noise-resilient mean-square consensus control is proposed by employing the proposed discrete-time distributed communication, which requires only to be implemented on local DG controllers, and DGs merely need a partial and limited knowledge of the problem parameters and can perform only local measurements, which is suitable for the plug-and-play functionality and dispersed nature of DG units; thus it improves the system reliability and the efficiency of the information utilization.

The remaining part of this chapter is organized as follows. In Sect. 2.2, we formulate the DC microgrid control problem and give some preliminaries. Section 2.3 gives the proposed discrete-time distributed mean-square consensus control scheme for achieving DC bus voltage restoration and maintaining proportional current sharing among all DGs by employing a sparse communication network subject to multiplicative noise and switching dynamic communication topology disturbances. Furthermore, the stability analysis for the whole closed-loop system is studied, and the control gains are designed. In Sect. 2.4, illustrative simulations are conducted on a DC microgrid to validate the effectiveness of the proposed controls. Section 2.5 draws the conclusions of this chapter.

Notation: Let $|\cdot|$ be the Euclidean norm. Denote $\mathbb{R}$, $\mathbb{N}$ as the set of real and natural numbers. If A is a vector or a matrix, then A^T denotes its transposes. If A is a matrix, denote by $\|A\|$ the operator norm of A, i.e., $\|A\| = \sup\{|Ax| : |x| = 1\}$. Moreover, let $\lambda_i(A)$ be the ith smallest eigenvalues of symmetric matrix A. Denote $\overline{A}$ as $\overline{A} = (A + A^T)/2$. Ω represents the set of events. The σ-algebra $\sigma\{\{\varsigma_l \in C\}, C \in \mathcal{C}, l = 1, 2, \ldots\}$ is denoted by F. P is used to denote the probability measure. For a given random variable ς, its expectation is denoted by $E(\varsigma)$. For a family of random variables $\{\varsigma_l, l = 1, 2, \ldots\}$, the σ-algebra is denoted by $\sigma\{\varsigma_l, l = 1, 2, \ldots\}$, where $\mathcal{C}$ is a Borel set. We say that a random variable is adapted to a σ-algebra F if ς is F-measurable.

2.2 Problem Formation and Preliminaries

The proposed distributed mean-square consensus cooperation control scheme for an islanded DC microgrid is shown in Fig. 2.1, where the low layer is the physical network and the high layer is a multi-agent cyber network. This architecture is able to assists the designed discrete-time distributed controllers under switching dynamic communication topology and multiplicative noise disturbances to achieve better performance. In the low physical level, the local primary control is employed to maintain the stability operation of physical nodes (e.g., DGs and loads). In the high cyber level, each cyber node is considered as an agent, which has the cooperative

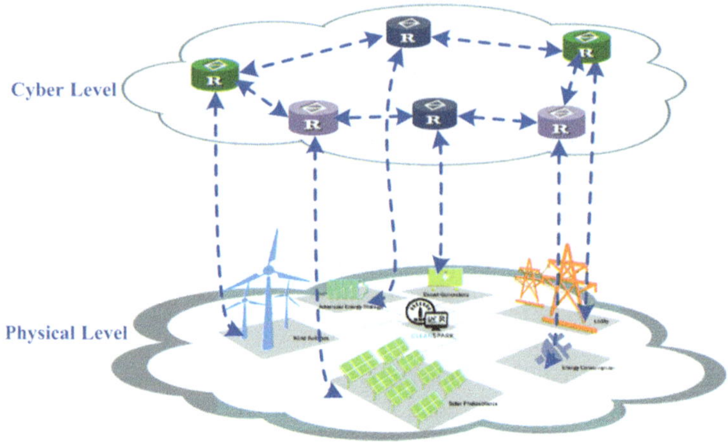

Fig. 2.1 A bilevel cooperation architecture for a DC microgrid, which is composed of physical and multi-agent cyber networks subjected to switching topology and multiplicative noise disturbances

communication and computational capabilities. Herein, in order to make our proposed control scheme more meaningful yet practical, the multi-agent cyber network is considered to suffer from switching dynamic communication topology and multiplicative noise disturbances.

2.2.1 Communication Topology in Cyber Layer

In this chapter, the required sparse communication network for DC microgrids can be modeled by a digraph $\mathcal{G}(\mathcal{V}, \varepsilon, \mathcal{A})$, where $\mathcal{V} = \{v_1, v_2, \cdots, v_N\}$ is a set of DG nodes and $\varepsilon \subseteq \mathcal{V} \times \mathcal{V}$ stands for a set of edges. The adjacency matrix is defined as $\mathcal{A}_{\mathcal{G}} = (a_{ij})_{N \times N}$ where $a_{ij} > 0$ if DG_i receives information from DG_j; otherwise, $a_{ii} = 0$. Also, the DG_j is called the neighbor of DG_i if $(v_i, v_j) \in \varepsilon$, and the set of neighbors for DG_i is defined as $\mathcal{N}_i = v_j \in \mathcal{V}$. The Laplacian matrix is defined as $\mathcal{L}_{\mathcal{G}} = \mathcal{D}_{\mathcal{G}} - \mathcal{A}_{\mathcal{G}}$, where $\mathcal{D}_{\mathcal{G}} = \text{diag}\{d_1, \cdots, d_N\}$ is the input degree matrix with $d_i = \sum_{j \in \mathcal{N}_i} a_{ij}$. The Laplacian matrix has all row sums equal to zero, that is, $\mathcal{L}_{\mathcal{G}} 1_N = 0$, with $1_N = (1, \cdots, 1)^T \in \mathbb{R}^N$. The union of $k$digraphs $G_1(V, \varepsilon_{G_1}, A_{G_1}), \ldots, G_k(V, \varepsilon_{G_k}, A_{G_k})$is defined as $\cup_{j=1}^{k} G_j = \left\{V, \cup_{j=1}^{k} \varepsilon_{G_j}, \cup_{j=1}^{k} A_{G_j}\right\}$. From the definitions of union digraph and Laplacian matrix, it is easy to see that $L_{\cup_{j=1}^{k} G_j} = \sum_{j=1}^{k} L_{G_j}$.

A directed graph has a spanning tree if the called root DG has at least input degree equal to one. The digraph $\overline{\mathcal{G}}$ is used to describe the interconnection topology of a microgrid consisting of one virtual leader-DG, denoted by 0, and follower DGs, denoted by $1, \ldots, N$. Define $\mathcal{B} = \text{diag}\{a_{10}, \ldots, a_{N0}\}$, where $a_{i0} > 0$ if DG_i receives messages from the leader; otherwise, $a_{i0} = 0$. Bear it in mind that the graph $\mathcal{G}$ is undirected; then $\mathcal{L} + \mathcal{B}$ is symmetrical, and all the eigenvalues of the matrix $\mathcal{L} + \mathcal{B}$ are positive (i.e., eigenvalues $0 < \lambda_1 \leq \lambda_2 \leq \cdots \leq \lambda_N$). In the following, some lemmas are given for the proof of the main results.

Lemma 1 ([24]) Let $(\varsigma_l)_{l \in N}$ be a martingale adapted to the filtration $(F_l)_{l \in N}$ on the probability space (Ω, F, P), and $\sup\limits_{l \in N} E\left(\|\varsigma_l\|^2\right) < \infty$. Then, this martingale converges to a random variable in mean-square sense.

Lemma 2 ([25]) $\lambda_i(\overline{A}) \leq \|A\|, i = 1, \ldots, N$ holds for any matrix $A \in \mathbb{R}^{N \times N}$.

Lemma 3 ([26]) Suppose that $\phi(k+n) \leq \sum\limits_{i=1}^{n} a_i \phi(k+i-1)$, where $\phi(i) \leq \sum 0$, $a(i) \leq \sum 0$, $i = 1, \ldots, n$, and $\gamma = \sum\limits_{i=1}^{n} a_i < 1$; then $\phi(k) \to 0$ as $t \to \infty$ and $\sum\limits_{k=1}^{\infty} \phi(k) < \frac{n}{1-\gamma} \max\{\phi(1), \ldots, \phi(n)\}$.

2.2.2 *Primary and Distributed Control of DGs in Physical Layer*

The detailed dynamical model based on local primary control of a DC microgrid will be discussed in this subsection. As represented in Fig. 2.2, a DC/DC buck converter-based DG unit is shown to supply power into the microgrid system. The primary control is a local control with droop characteristics, which employs to realize voltage and current management. In particular, comparing the output voltage of DG_i with the nominal set points v_i^{nom} controlled by droop control in the voltage control loop, the voltage errors are calculated to supply the reference values i_i^{ref}. By employing the reference values provided from the voltage loop, the current errors are calculated and finally utilized to manage the duty cycle d_i^{buck} of the inverter outputs by pulse width

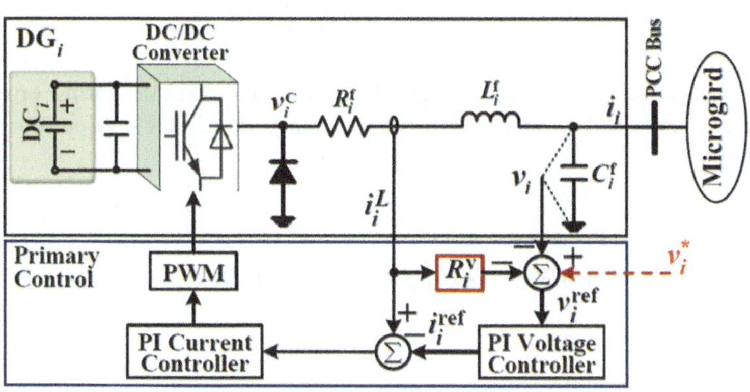

Fig. 2.2 A general primary and distributed control for the DG_i in a DC microgrid

modulation (PWM) mode. The following proportional integral (PI) controllers are employed in inner voltage and current control loops of each DG unit. The droop-based local primary control for DG_i can be described as

$$v_i = v_i^* - R_i^{\mathrm{v}} i_i \tag{2.1}$$

where v_i^*, i_i are the nominal set point voltage and measured current of DG_i, respectively, and R_i^{v} is the virtual resistance. It can be calculated by $R_i^{\mathrm{v}} = \Delta v_i / i_{i\max}$, where Δv_i and $i_{i\max}$ are the maximum allowed voltage deviation and the maximum output current, respectively.

The dynamic of local droop-based primary control consisting of PI voltage/current control loops is depicted as [15]

$$\begin{aligned}
\frac{\mathrm{d}\alpha}{\mathrm{d}t} &= v_i^* - R_i^{\mathrm{v}} i_i^L - v_i^{\mathrm{C}}, \\
i_i^{\mathrm{ref}} &= \kappa_P^v \left(v_i^* - R_i^{\mathrm{v}} i_i^L - v_i^{\mathrm{C}}\right) + \kappa_I^v \alpha, \\
\frac{\mathrm{d}\beta}{\mathrm{d}t} &= i_i^{\mathrm{ref}} - i_i^L, \\
d_i^{\mathrm{buck}} &= \kappa_P^i \left(i_i^{\mathrm{ref}} - i_i^L\right) + \kappa_I^i \beta,
\end{aligned} \tag{2.2}$$

where α and β are auxiliary variables and κ_P^v, κ_I^v and κ_P^i, κ_I^i are the coefficient terms of PI voltage and current controllers, respectively. v_i^{C} and i_i^L represent the capacitance voltage and inductor current of LC filter, respectively; d_i^{buck} is the duty cycle of the buck converter i; R_i^{f}, L_i^{f}, and C_i^{f} are the filter resistance, inductor, and capacitor, respectively.

The distributed cooperative control is responsible for mitigating the undesired voltage deviation along with load current sharing. Specifically, by utilizing DG_i's information exchange from itself and its neighbors, the objective can be achieved through adjusting the local voltage state set point value v_i^{nom} for each DG local control v_i^* as shown in Fig. 2.2, which can be expressed as

$$v_i^* = v_i^{\mathrm{nom}} + \Delta v_i + \Delta R_i^{\mathrm{v}} i_i, \tag{2.3}$$

where Δv_i and $\Delta R_i^{\mathrm{v}} i_i$ are the voltage shift term and current sharing error term, respectively, obtained from the distributed voltage controller and current sharing controller.

2.2.3 *Primary and Distributed Control of DGs in Physical Layer*

The objective of the proposed control scheme for DC microgrids is to achieve DC bus voltage restoration and maintain proportional current sharing among massive

DGs in mean square. Furthermore, we consider the communication network suffering from switching dynamic communication topology and multiplicative noise disturbances. In this section, the discrete-time distributed voltage control employing a leader-follower consensus protocol will be discussed, and discrete-time distributed current sharing control employing a leader-less consensus protocol will be discussed in the following subsection. The proposed distributed cooperative control algorithms select the voltage and current control inputs, u_i^v and u_i^i for the voltage and current of each DG$_i$'s terminal, v_i and i_i, in order to update the voltage set points v_i^* in (2.3). Accordingly, the terminal voltage v_i of each DG unit is able to be restored to its desired value v^{ref}. When designing a discrete-time distributed control scheme for islanded microgrids, utilization of communication networks for information exchange among neighboring DGs will introduce the following fundamental issues.

Noise Disturbances Note that communication noise disturbances are inherent in a communication system, which are usually caused by limited network bandwidth or congested network traffic. As a result, each DG in a microgrid can only receive the noise-dependent information from its neighbors, and the information transmission during any neighbors may experience different communication noises. The adverse effects of communication noise can be categorized as (*i*) state-independent additive noises addressed in [16, 18] and (*ii*) multiplicative state-dependent noises addressed in our work. As usual, the noise processes $\{\xi_{ij}(t) : i,j = 1,\ldots,N\}$ are assumed as independent standard white noises, that is,

$$\int_0^t \xi_{ij}(s)ds = \varpi_{ij}(t), t \geq 0, \tag{2.4}$$

where $\varpi_{ij}(t)$ are independent Brownian motions and ξ_{ij} represents the noise intensity. Eventually, this results in the whole information exchange disturbed by communication noises. Furthermore, $\{\xi_{ji}(k), k \in \mathbb{N}, j \in \{0\} \cup \mathcal{V}, i \in \mathcal{V}\}$ are assumed to satisfy the following assumption:

Assumption 2.1 $\{\xi_{ji}(k), k \in \mathbb{N}, j \in \{0\} \cup \mathcal{V}, i \in \mathcal{V}\} \subset \psi$ where
$\psi = \{\varsigma \mid (\varsigma(l) \in, F^{\varsigma}(l))$ is a martingale difference

$$\sigma_{\varsigma} \triangleq \sup_{l \geq 0} E\|\varsigma(l)\|^2 < \infty$$

with $F^{\varsigma}(l) = \sigma\{\varsigma(0),\ldots,\varsigma(l)\}$. $\xi_{ji}(k), i \in \{0\} \cup \mathcal{V}, j \in \mathcal{V}, l \in \mathbb{N}$ are mutually independent and also independent of the initial conditions $v_i(0)$ and v^{ref}.

Switching Dynamic Communication Topologies As stated in Sect. 2.1, communication topologies of microgrids may be frequently changed according to specific communication requirements or environmental disturbances. Thus, robustness against microgrid topological uncertainties and maintaining the plug-and-play functionality of DG units is another property that is integrated with the proposed robust control scheme. Since the plug-in (out) of DGj to (from) DGi affects the adjacency

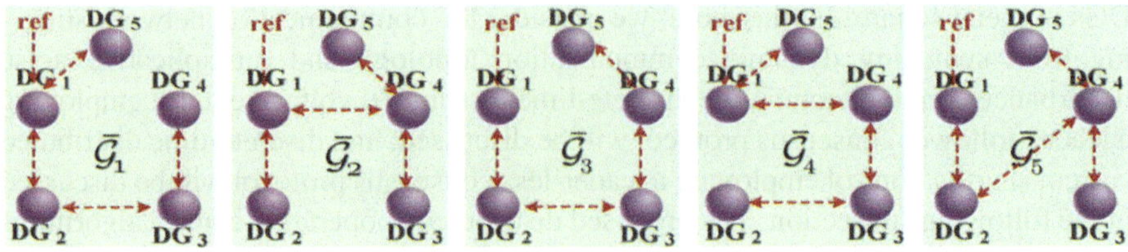

Fig. 2.3 Switching dynamic communication topologies across a DC microgrid

matrix $\mathcal{A}_{\mathcal{G}}$, taking the following case, for example, the communication topology for DGs is originally in a pattern as illustrated in Fig. 2.3. Due to environmental disturbances, two communication links (between DGs 2 and 3, DGs 1 and 5) are failed for a while. Then, other backup communication links (between DGs 1 and 4, DGs 4 and 5) are built; that is, the communication topology is transformed into a pattern with $\overline{\mathcal{G}}_2$. When another disturbance occurs, the topology may be changed to another pattern. As a result, the communication topology can be switched among four patterns shown in Fig. 2.3. Due to possible intermittent communications with the reference DG0, the following assumption is introduced:

Assumption 2.2 There is an $l > 0$ such that in every time interval $[k, k + l - 1]$, $k \in \mathbb{N}$, the group of digraphs $\{\mathcal{G}(k), \ldots, \mathcal{G}(k+l-1)\}$ jointly contains a spanning tree and $0 < \left\| \sum_{i=k}^{k+l-1} \mathcal{B}(k) \right\| < \infty$. Furthermore, $\inf_{t\to\infty} a_{ij}(k) > 0$ for all $a_{ij}(k) \neq 0$. Assumption 2 implies that $\inf_{k\geq 0} \lambda_k^l > 0$.

Two control objectives are considered in this chapter for DC microgrids, i.e., voltage restoration and proportional current sharing in mean square due to the influence of switching topology and multiplicative noise disturbances. More specifically, the proposed distributed cooperative control scheme aims to select the nominal voltage set point v_i^* in (2.3) to restore the terminal voltage v_i of each DGi to its desired value v^{ref} supplied by the virtual-leader DG0 in mean square, i.e.,

$$\lim_{k\to+\infty} E\left\| v_i(k) - v^{\text{ref}} \right\|^2 = 0, \tag{2.5}$$

for $i = 1, \cdots, N$, where $v_i(k)$ is the measured voltage of DGi at time k, $k \in \mathbb{N}$; meanwhile, share the overall load currents in mean square according to their respective capacities, i.e.,

$$\lim_{k\to+\infty} E\left\| \frac{i_i(k)}{i_{i\max}} - \frac{i_j(k)}{i_{j\max}} \right\|^2 = 0, \tag{2.6}$$

for all $i \neq j \in \{1, 2, \ldots, N\}$, where $i_i(k)$ and $i_j(k)$ represent the output currents of DGi and DGj at time k, respectively. $i_{i\max}$ and $i_{j\max}$ represent the capacity (maximum output current) of DGi and DGj at time k, respectively. Since different DGs have different capacities in a DC microgrid, it is desirable to achieve proportional load current sharing according to their respective capacities. As a result, the utilization level of different DGs can be identical in steady states.

2.3 Distributed Mean-Square Consensus for DC Microgrids Under Switching Topology and Multiplicative Noises

The network controlled DC microgrid is a cyber-physical system, where the electrical system and communication system are coupled with closed-loop feedback control. To ensure the DC bus voltage restoration and proportional current sharing among DGs, a discrete-time distributed mean-square consensus cooperative control scheme will be developed, which is fully distributed and robust to switching dynamic communication topology and multiplicative noise disturbances. Specifically, the proposed distributed state-dependent noise-resilient cooperative controllers require only to be implemented on local DG controllers, and DGs merely need a partial and limited knowledge of the problem parameters and can perform only local measurements, which is suitable for the plug-and-play feature and dispersed nature of DG units; thus, it improves the system reliability and the efficiency of the information utilization.

2.3.1 Distributed Voltage Restoration

To facilitate the analysis with switching topologies for the variable communication network, a switching indicator $\sigma(t) : [0, +\infty) \rightarrow \rho_\Gamma = \{1, 2, \ldots, m\}$ is employed to determine the graph characteristics at a particular time, which is defined on digraphs $\Gamma = \{\overline{\mathcal{G}}(k), k \in \mathbb{N}$, where $\overline{\mathcal{G}}(k)$ is a balanced digraph.

Since all DGs are supposed to merely communicate with their neighbors through a sparse network subjected to switching topology and multiplicative noise disturbances, for $\mathcal{G}(k) \in \Gamma$, the discrete-time system states are updated as

$$v_i(k+1) = v_i(k) + u_i^v(k) \tag{2.7}$$

where $u_i^v(k)$ is the distributed leader-follower mean-square consensus voltage controllers, which can be designed as

$$u_i^v(k) = c\sum_{j\in\mathcal{N}_i} a_{ij}(k)\left[\varpi_{ji}^v(k) - v_i(k)\right] + a_{i0}(k)\left[\varpi_{i0}^v - v_i(k)\right] \tag{2.8}$$

such that $\lim_{k\to+\infty} E\left\|v_i(k) - v^{\text{ref}}\right\|^2 = 0$ for $i = 1, \cdots, N$ and $k \in \mathbb{N}$, where $\varpi_{ji}^v(k) = v_j(k) + \mid v_j(k) - v_i(k) \mid \xi_{ji}(k)$ and $\varpi_{i0}^v(k) = v_0^{\text{ref}} + \mid v_0^{\text{ref}} - v_i(k) \mid \xi_{i0}(t)$ respectively represent the state information received from DGi's neighbors and accessed the voltage reference v^{ref} corrupted by multiplicative noises. $c > 0$ is the control gain to be designed. $\mathcal{A}_k = \left(a_{ij}(k)\right)_{N\times N}$ is the adjacency matrix of DGi, corresponding to the communication network $\mathcal{G}$. $\mathcal{B}(k) =$ diag $\{a_{10}(k), \ldots, a_{N0}(k)\}$. DG$i$ can access the voltage reference v^{ref} if and only if $a_{i0}(k) > 0$.

2.3.2 Distributed Current Sharing Consensusablity

Since different DGs have different capacities in a microgrid, it is desirable to achieve proportional load current sharing according to their respective capacities. As a result, the utilization level of different DGs can be identical in steady states [12, 15], i.e.,

$$\frac{i_1}{i_{1\max}} = \ldots = \frac{i_i}{i_{i\max}} = \ldots = \frac{i_N}{i_{N\max}}. \tag{2.9}$$

By the above preparations, we can now design a consensus-based distributed control protocol to ensure Eq. (2.9) holds. Based on the distributed communication network subjected to switching dynamic communication topology and multiplicative noise disturbances, each local controller can exchange its local current loading percentage $\frac{i_i}{i_{i\max}}$ with its neighbors. To this end, taking the time derivative on both sides of (2.9) yields

$$\frac{i_i(k+1)}{i_{i\max}} = \frac{i_i(k)}{i_{i\max}} + u_i^i(k), \tag{2.10}$$

where the mean-square consensus current sharing controllers $u_i^i(k)$ can be designed as

$$u_i^i(k) = c\sum_{j\in\mathcal{N}_i} a_{ij}(k)\left[\varpi_{ji}^i(k) - \frac{i_i(k)}{i_{i\max}}\right], \tag{2.11}$$

such that $\lim_{k\to+\infty} E\left\|\frac{i_i(k)}{i_{i\max}} - \frac{i_j(k)}{i_{j\max}}\right\|^2 = 0$ for all $i \neq j \in \{1, 2, \ldots, N\}$ and $k \in \mathbb{N}$, where $\varpi_{ji}^i(k) = i_j(k) + \mid i_j(k) - i_i(k) \mid \xi_{ji}(k)$represents the state information received from DGi's neighbors corrupted by multiplicative noises. $c > 0$ is the control gain to be designed. $\mathcal{A}_k = \left(a_{ij}(k)\right)_{N\times N}$ is the adjacency matrix of DGi, corresponding to the communication network $\mathcal{G}$.

Now with the above load current sharing control inputs $u_i^i(k)$ and the previous voltage control inputs $u_i^v(k)$, for all $k \in \mathbb{N}$, the nominal set points $v_i^*(k)$ utilized in the primary control procedure can be calculated as

$$v_i^*(k) = \int \left[u_i^v(k) + R_i^v u_i^i(k)\right] dt. \tag{2.12}$$

2.3.3 Stability Analysis of the Closed-Loop System

In the following, the graph theory, the stochastic theory, and the Lyapunov technique are employed to derive some stability criteria to guarantee feasibility of the closed-loop microgrid system.

Denote $v(k) = (v_1(k), v_2(k), \ldots, v_N(k))^T \in \mathbb{R}^N$, $v^{\text{ref}} = 1_N \bigotimes v^{\text{ref}} \in$

$$\begin{aligned}\mathbb{R}^N, \varpi^v(k) &= \text{diag}\{\varpi_1(k), \varpi_2(k), \ldots, \varpi_N(k)\}, \varpi_i^v(k)\\ &= \text{diag}\{|v_1(k) - v_i(k)|, |v_2(k) - v_i(k)|, \ldots, |v_N(k) - v_i(k)|\}\end{aligned}$$

for $i = 1, \ldots, N, \xi(k) = \left(\xi_1^T(k), \xi_2^T(k), \ldots, \xi_N^T(k)\right)^T$, $\xi_i^T(k) = (\xi_{1i}(k), \xi_{2i}(k), \ldots, \xi_{Ni}(k))^T$, $\varpi_0^v(k) = \text{diag}\{|v_0(k) - v_1(k)|, |v_0(k) - v_2(k)|, \ldots, |v_0(k) - v_N(k)|\}$ and $\xi_0(k) = (\xi_{01}(k), \xi_{02}(k), \ldots, \xi_{0N}(k))^T$. $\Pi(h, h) = I_N$, $\Pi(n + 1, h) = (I_N - cH(n))\Pi(n, h)$, $h = 0, \ldots, n$, $n \in Z^+$, $\xi_\varsigma^\tau = \sum_{j=\varsigma}^{\varsigma+\tau-1}\Pi(\varsigma+\tau-1, j)c\mathcal{D}_{\mathcal{G}(j)}\varpi(j)^v\xi(j)$, and $\xi_{0\varsigma}^\tau = \sum_{j=\varsigma}^{\varsigma+\tau-1}\Pi(\varsigma+\tau-1, j)c\mathscr{B}(j)\varpi_0^v(j)\xi_0(j)$, $H(k) = \mathcal{L}_{\mathcal{G}(k)} + \mathscr{B}(k)$, $\overline{H}_k^l = \sum_{i=k}^{k+l-1}\overline{H}(i)$, $\lambda_k^l = \lambda_1\left(\overline{H}_k^l\right)$, $\Xi_l = \rho_l \max\left\{\sup_{k\geq 0}\|H(k)\|, 1\right\}^{2l}$ where $\rho_l = 2^{2l} - 2l - 1$, $\lambda_k^l = \lambda_1\left(H_{\overline{\mathcal{G}}_k^l}\right)$.

With these denotations, substituting protocol (2.8) into (2.7), the closed loop system can be rewritten in a compact form:

$$\begin{aligned}v(k+1) = (I_N - cH(k))v(k) + c\mathcal{D}_{G(k)}\varpi^v(k)\xi(k)\\ +c\mathscr{B}(t)\varpi_0^v(k)\xi_0(k),\end{aligned} \tag{2.13}$$

where $\mathcal{D}_{G(k)} = \text{diag}\{a(1,\cdot)(k), \ldots, a(N,\cdot)(k)\}$ is an $N \times N^2$ block diagonal matrix with $a(i, \cdot)(k)$ being the ith row of the adjacency matrix $A(k) = (a_{ij}(k))_{N\times N}$ of the digraph $\mathcal{G}$. Because the digraph $\mathcal{G}(k)$ is balanced, we have $1^T L_G = 0$ and $L_G 1 = 0$. By defining $e(k) = v(k) - v^{\text{ref}}$ and choosing $c \leq \frac{1}{\Omega}$, we have

$$e(k+1) = (I_N - cH(k))e(k) + c\mathcal{D}_{G(k)}\varpi^{v}(k)\xi(k) + c\mathcal{B}(k)\varpi_0^{v}(k)\xi_0(k), \tag{2.14}$$

and hence,

$$e(k+l) = \Pi(k+l,t)e(t) + \xi_k^l + \xi_{0k}^l. \tag{2.15}$$

By some simple calculations, we have

$$\left\| \Pi^T(k+l,t)\Pi(k+l,t) - I_N + 2c\left[\sum_{s=k}^{k+l-1} \overline{H}_{\mathcal{G}(s)}\right] \right\| \le c^2\Xi_l. \tag{2.16}$$

Then by considering the Lyapunov candidate as $V(v(k)) = e(k)^T e(k)$ and (2.16), we obtain

$$\begin{aligned} V(v(k+l)) \le V(v(k)) &- 2ce^T(k)\left[\sum_{s=k}^{k+l-1} \overline{H}_{G(s)}\right]e(k) \\ &+ \left(\xi_k^l\right)^T \xi_k^l + \left(\xi_{0k}^l\right)^T \xi_{0k}^l \\ &+ 2e^T(k)\Pi(k+l,k)\xi_k^l \\ &+ c^2\Xi_l V(v(k)), \end{aligned} \tag{2.17}$$

Furthermore, let $E\left[\xi_{ij}^2(k)\right] = \sigma_{ij}^2(k)$ and $\overline{\sigma}^2 = \sup_{k\ge 0}\left\{a_{ij}^2\sigma_{ij}^2(k) | i,j \in V\right\}$, $E\left[\xi_{0i}^2(k)\right] = \sigma_{0i}^2(k)$ and $\overline{\sigma}_0^2 = \sup_{k\ge 0}\left\{a_{0i}^2\sigma_{0i}^2(k) | i \in V\right\}$; we obtain

$$E\left[\left(\xi_k^l\right)^T \xi_k^l\right] \le 4c^2\overline{\sigma}^2 N\sqrt{N}\sum_{i=0}^{l-1} E[V(v(k+i))], \tag{2.18}$$

and

$$E\left[\left(\xi_{0k}^l\right)^T \xi_{0k}^l\right] \le k^2\overline{\sigma}_0^2\sqrt{N}\sum_{i=0}^{l-1} E[V(v(k+i))]. \tag{2.19}$$

Because the digraph $\mathcal{G}(k)$ is balanced, we know that $H(t)$ is the Laplacian matrix of the symmetrized graph $\mathcal{G}(k)$ of $\mathcal{G}$. Because the digraph $\mathcal{G}(k)$ contains a spanning directed tree, we know that $\overline{\mathcal{G}}$ is connected and $\lambda_1\left(\overline{H}\right)$. Therefore, from Lemma 1 and (2.17), noticing that $\sum_{s=k}^{k+l-1} \overline{H}_{\mathcal{G}(s)} = H_{\overline{\mathcal{G}}_k^l}$, by taking expectation on both sides of the above inequality and considering Assumption 1, we have

$$
\begin{aligned}
E[V(v(k+l))] \leq & (1-2c\lambda_k^l + c^2\Xi_l)E[V(v(k))] \\
& + c^2\sqrt{N}(4N\overline{\sigma}^2 + \overline{\sigma}_0^2)\sum_{i=0}^{l-1} E[V(v(k+i))] \\
= & c^2\sqrt{N}(4N\overline{\sigma}^2 + \overline{\sigma}_0^2)\sum_{i=1}^{l-1} E[V(v(k+i))] \\
& + \theta E[V(v(k))],
\end{aligned}
\tag{2.20}
$$

where $\theta = 1 - 2c\inf_{t\geq 0}\lambda_k^l + c^2\Xi_l + c^2\sqrt{N}(4N\overline{\sigma}^2 + \overline{\sigma}_0^2)$.

Note that $\rho_l \geq l^2$ and $\max\left\{\sup_{k\geq 0}\|H(k)\|, 1\right\}^{2l} \geq \|H(i)\|\|H(j)\|, i, j \in N$. We have

$$
\begin{aligned}
\Xi_l \geq l^2\left(\max\left\{\sup_{k\geq 0}\|H(k)\|, 1\right\}^{2l}\right) \geq \left\|\sum_{s=k}^{k+l-1} H_{\mathcal{G}(s)}\right\|^2 \\
\geq \lambda_1\left(\sum_{s=k}^{k+l-1}\overline{H}_{\mathcal{G}(s)}\right)^2 \geq \left(\inf_{t>0}\lambda_k^l\right)^2.
\end{aligned}
\tag{2.21}
$$

The second inequality is due to the reason that $\sum_{i=k}^{k+l-1}\sum_{j=k}^{k+l-1}\|H_{\mathcal{G}(i)}\| \times \|H_{G(j)}\| \geq \left\|\sum_{s=k}^{k+l-1} H_{\mathcal{G}(s)}\right\|^2$ and the third one comes from Lemma 2. From Lemma 3, it is easy to check that if c satisfies $0 < c < \min\left\{\frac{1}{\Omega}, \frac{2\inf_{k\geq 0}\lambda_k^l}{\Xi_l + l\sqrt{N}(4N\overline{\sigma}^2 + \overline{\sigma}_0^2)}\right\}$, then $0 < \theta < 1$.

It follows from $E[V(v(k))] \to 0$ as $k \to \infty$ that $\lim_{k\to\infty} E\|v(k) - v^{\text{ref}}\|^2 = 0$ for $i = 1$, ..., N. The proof is thus completed. Eventually, the conclusion of the proposed distributed controllers can be described in Theorem 2.1.

Theorem 2.1 Suppose that each balanced digraph $\overline{\mathcal{G}}_{\sigma(k)}$ contains a spanning tree with a root located in the virtual reference DG0 for any switching signal $\sigma(k) \in \rho_\Gamma$. If the control gain c satisfies $0 < c < \min\left\{\frac{1}{\Omega}, \frac{2\inf_{k\geq 0}\lambda_k^l}{\Xi_l + l\sqrt{N}(4N\overline{\sigma}^2 + \overline{\sigma}_0^2)}\right\}$, where $\Omega = \sup_{k\geq 0}\max{}_i(d_i(k) + a_{i0}(k))$, then the distributed voltage controllers (2.8) and distributed current sharing controllers (2.11) can restore the DG output voltage v_i to the reference state v^{ref} and maintain proportional current sharing among DGs in mean square, even if the sparse communication network suffers from switching topology and multiplicative noise disturbances.

Compared to $u_i^v(k)$ and $u_i^i(k)$ in (2.8) and (2.11), there is no second term with pinning gain $a_{i0}(k)$ to a reference value. Therefore, we can modify Theorem 2.1 to prove that the proposed control protocol in (2.11) satisfies (2.8) without a specific final synchronization term.

2.4 Controller Performance Validation

To demonstrate the performance of the proposed control scheme, the test DC microgrid with the same topology introduced in [15] as shown in Fig. 2.4 is employed to be implemented in MATLAB/SimPowerSystems. Other used parameters of the test microgrid are shown in Table 2.1. Taking the switching topology $\overline{\mathcal{G}}_{\sigma(k)} = \{\overline{\mathcal{G}}_1, \ldots, \overline{\mathcal{G}}_5\}$, the controlled DC microgrid begins at $\overline{\mathcal{G}}_1$ and switches in a random dynamic order among five networks as shown in Fig. 2.3. The associated adjacency matrix of $\overline{\mathcal{G}}_1$ is

$$A = \begin{pmatrix} 0 & 1 & 0 & 0 & 1 \\ 1 & 0 & 1 & 0 & 0 \\ 0 & 1 & 0 & 1 & 0 \\ 0 & 0 & 1 & 0 & 1 \\ 1 & 0 & 0 & 1 & 0 \end{pmatrix}.$$

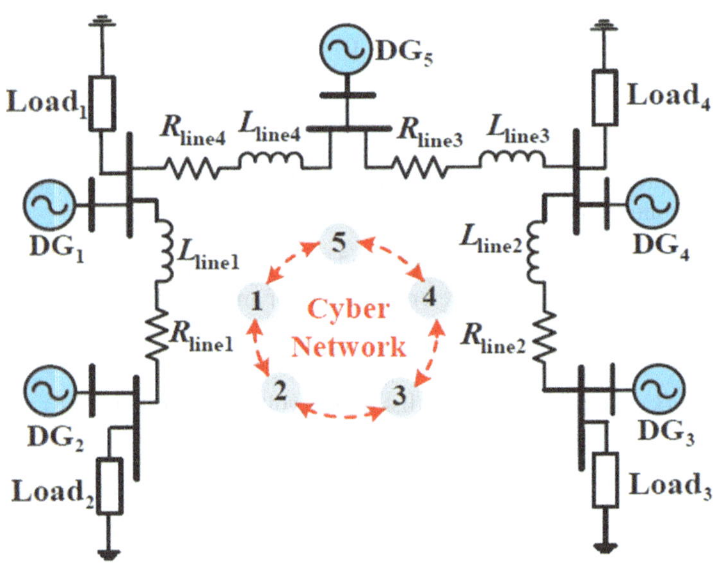

Fig. 2.4 The test DC microgrid with DGs and local loads ("IEEE from [15]")

Table 2.1 Parameters for the test DC microgrid system

DG_1 & DG_4		DG_2 & DG_3 & DG_5	
V_{DC}	100 V	V_{DC}	80 V
L^f	2 mH	L^f	1.95 mH
L^o	2 mH	L^o	1.95 mH
C^f	1.6 μF	C^f	1.5 μF
R^v	0.5 Ω	R^v	0.2 Ω
K_v^{PI}	5/560	K_v^{PI}	4/800
K_i^{PI}	1.2/97	K_i^{PI}	1/97
$Load_1$	$Load_2$	$Load_3$	$Load_4$
11 Ω	17 Ω	13 Ω	15 Ω
$Line_1$	$Line_2$	$Line_3$	$Line_4$
0.05 Ω 3 μH	0.1 Ω 1.5 μH	0.05 Ω 3 μH	0.1 Ω 1.5 μH

ξ_{ij} are Gaussian white noises with distribution $N(0, 3)$ in the whole communication process; then we take the control gain $c = 0.5$.

2.4.1 *Performance Assessment with Complex Operation Ability*

In this case, the performance of the proposed noise-resiliency controllers in switching dynamic communication topology $\overline{\mathcal{G}}_{\sigma(k)}$ is designed in the situation of five stages as follows:

1. At $t = 0$ s. The proposed voltage controller (2.8) is activated.
2. At $t = 5$ s. The proposed current sharing controller (2.11) is applied.
3. At $t = 10$s. $Load_1$ and $Load_3$ are added.
4. At $t = 15$ s. $Load_1$ and $Load_3$ are removed.
5. At $t = 20$s. DG_1 and DG_2 are plugged out.

As shown in Fig. 2.5, at $t = 0$ s, only the proposed voltage controller (2.8) is activated to recover the terminal voltage to the reference $v^{ref} = 48V$. However, the current sharing cannot be achieved (see Fig. 2.5b). Subsequently, the voltage restoration and current sharing are always realized no matter the load is or not varied. At $t = 20$s, due to DG_1 and DG_2 plugged out from the system, their output currents are restored to their initial state invalid for the proposed control algorithm. In turn, the remaining DGs achieve new proportional current sharing by exchanging the neighbor's shared information. The interaction during the whole process goes smoothly since DG can respond quickly, causing the proposed noise-resiliency distributed controllers to enable DGs to reach the steady state rapidly. As seen in Fig. 2.5a, b, since the communication topology switches in a random order among five networks as shown in Fig. 2.3, the output voltage and current express a slight but acceptable jitter phenomenon as shown in the zoomed version of Fig. 2.5. Overall,

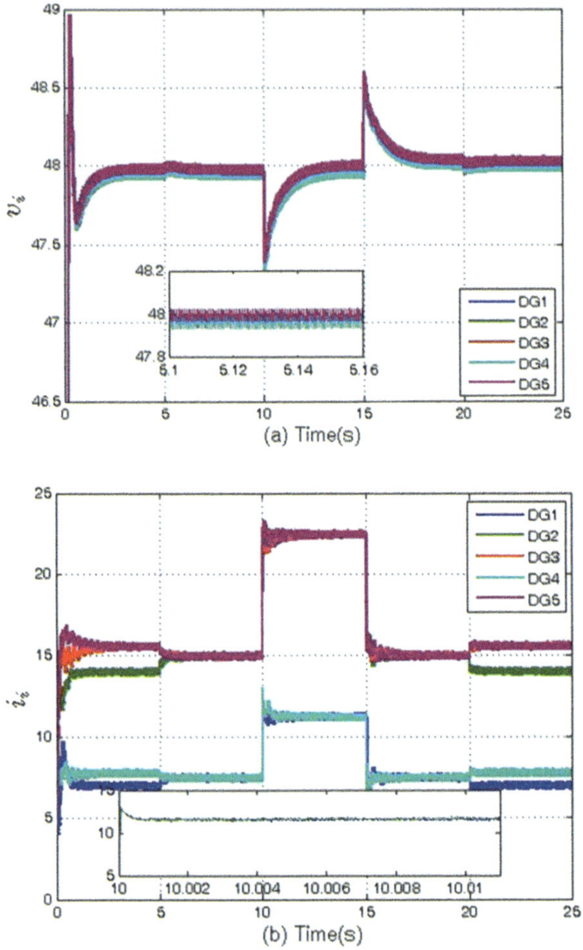

Fig. 2.5 System performance of the proposed noise-resiliency distributed control with multiplicative noise: (**a**) terminal voltage; (**b**) current output

Fig. 2.5 shows that the designed controllers can realize the voltage restoration and proportional current sharing among DGs, even if the communication network is subject to switching topology and multiplicative noise disturbances.

2.4.2 *Performance Assessment with Larger Noise Disturbances*

In this case, the performance of the proposed noise-resiliency controllers with larger noise disturbances considering distribution $N(0, 6)$ is designed in the situation. It can be seen from Fig. 2.6 that even if the noise interference is doubled, the synchronous voltage and current sharing process of the whole system can still be well realized.

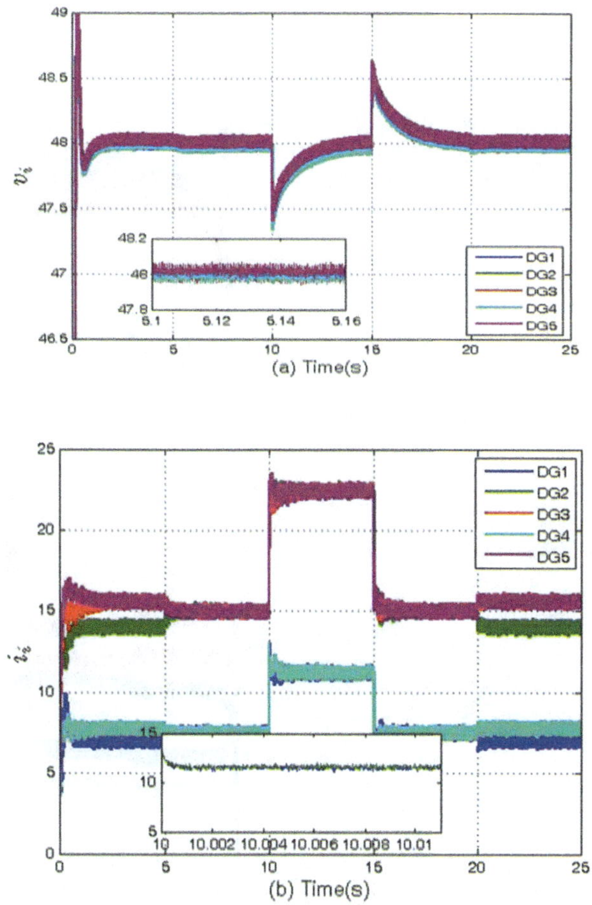

Fig. 2.6 System performance of the proposed noise-resiliency distributed control with larger multiplicative noise: (**a**) terminal voltage; (**b**) current output

2.4.3 Performance Comparative Analysis

To highlight the performance of the proposed state-dependent multiplicative noise-resiliency distributed control strategy, a comparative evaluation has been performed with the asymptotic controller proposed with fixed communication topology $\overline{\mathcal{G}}_1$ in [15]. We perform the simulation scenario of case A by exploiting the distributed controller proposed in [15] and present the results in Fig. 2.7. As shown in Fig. 2.7, it can be observed that when the state-dependent multiplicative noise is added with distribution $N(0, 3)$ at $t = 0$ s employing the previously proposed controllers in [15], it results in the larger overshoot on the output voltage and current sharing in transient, which destroys the synchronization performance of voltage and current sharing.

Furthermore, compared with Fig. 2.7, the convergence time of the system dynamic shown in Fig. 2.7 is prolonged by the noise disturbances. On the contrary,

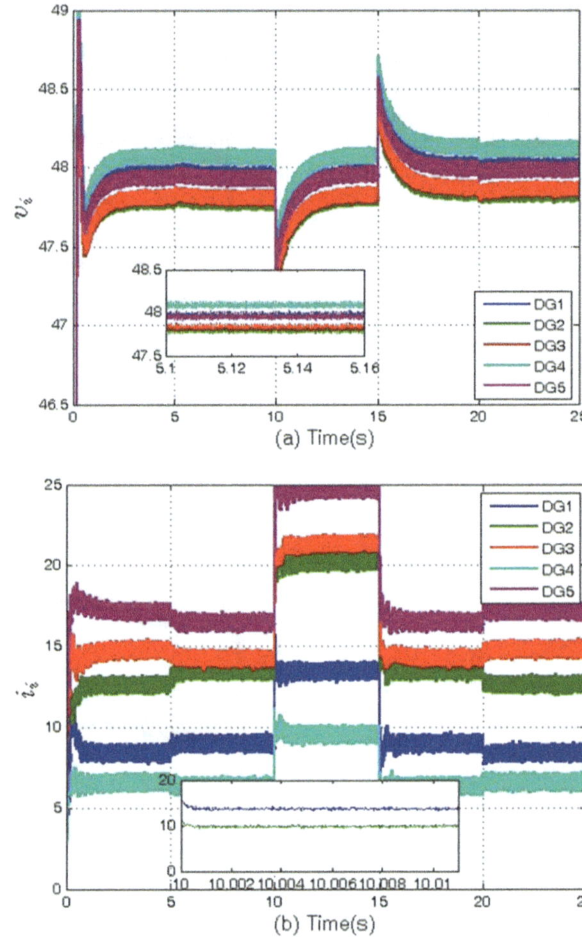

Fig. 2.7 Comparative effectiveness between the proposed controllers and the proposed controllers in [15] under state-dependent multiplicative noise: (**a**) terminal voltage; (**b**) current output

since the designed control gain c considering the noise term in Theorem 2.1, it satisfies attenuate the measurement noises. Eventually, the proposed noise-resiliency distributed controllers have desirable performance as illustrated in Fig. 2.5. Note that since the proposed controllers need to calculate the amount of noise related to information in the process of information interaction, its calculation amount will be larger than that of the traditional ideal state controller.

2.5 Conclusion

A discrete-time distributed bilevel cooperation scheme for DC microgrids has been investigated in this chapter, where the distributed communication network is subject to switching dynamic communication topology and multiplicative noise disturbances. The proposed controllers simultaneously achieve both accurate power sharing and voltage restoration through DGs operating collaboratively by employing merely neighbor to neighbor over a sparse communication network. Furthermore, by using the stochastic analysis techniques combined with the Lyapunov function, sufficient conditions on the control gain designed considering noise disturbances and switching dynamic communication topology are established to guarantee the stability and reliability of the whole systems. Case studies of five DGs are performed under switching topologies with multiplicative noise disturbances, which have shown the effectiveness of the proposed controllers.

References (IEEE Transaction PES)

1. J. Lai, X. Lu, X. Yu, A. Monti, Cluster-oriented distributed cooperative control for multiple AC microgrids. IEEE Trans. Ind. Inf. **15**(11), 5906–5918 (2019)
2. X. Lu, J. Lai, G.P. Liu, Master-slave cooperation for multi-dc-mgs via variable cyber networks. IEEE Trans. Cybernetics, to be published (2020). https://doi.org/10.1109/TCYB.2020.3035587.
3. A. Milczarek, M. Malinowski, J.M. Guerrero, Reactive power management in islanded microgrid--Proportional power sharing in hierarchical droop control. IEEE Trans. Smart Grid **6**(4), 1631–1638 (2015)
4. X. Lu, X. Yu, J. Lai, J.M. Guerrero, Y. Wang, A novel distributed secondary coordination control approach for islanded microgrids. IEEE Trans. Smart Grid **9**(4), 2726–2740 (2018)
5. M. Dong, L. Li, Y. Nie, D. Song, J. Yang, Stability analysis of a novel distributed secondary control considering communication delay in DC microgrids. IEEE Trans. Smart Grid **10**(6), 6690–6700 (Nov. 2019)
6. X. Lu, J. Lai, Two-layer cooperative control for multiple converter network clusters. IEEE Trans. Circuits Syst. II Exp. Briefs, to be published (2020). https://doi.org/10.1109/TCSII.2020.2971085.
7. J. Lai, X. Lu, X. Yu, A. Monti, Stochastic distributed secondary control for ac microgrids via event-triggered communication. IEEE Trans. Smart Grid **11**(4), 2746–2759 (2020)
8. J. Lai, X. Lu, Resilient distributed voltage synchronization of ci networks under denial of service attacks. IEEE Trans. Circuits Syst. II Exp. Briefs, to be published (2020). https://doi.org/10.1109/TCSII.2020.3040241.
9. X. Lu, J. Lai, X. Yu, A novel secondary power management strategy for multiple ac microgrids with cluster-oriented two-layer cooperative framework. IEEE Trans. Ind. Infor. **17**(2), 1483–1495 (2021)
10. D.H. Dam, H.H. Lee, A power distributed control method for proportional load power sharing and bus voltage restoration in a DC microgrid. IEEE Trans. Ind. Appl. **54**(4), 3616–3625 (2018)
11. P.H. Huang, P.C. Liu, W. Xiao, M.S. El Moursi, A novel droop-based average voltage sharing control strategy for DC microgrids. IEEE Trans. Smart Grid **6**(3), 1096–1106 (2015)

12. M. Tucci, L. Meng, J.M. Guerrero, G. Ferrari-Trecate, Stable current sharing and voltage balancing in DC microgrids: A consensus-based secondary control layer. Automatica **95**, 1–13 (2018)
13. M. Cucuzzella, S. Trip, et al., A robust consensus algorithm for current sharing and voltage regulation in dc mirogrids. IEEE Trans. Control Syst. **27**(4), 1583–1595 (2018)
14. D. Pullaguram, S. Mishra, N. Senroy, Event-triggered communication based distributed control scheme for DC microgrid. IEEE Trans. Power Syst. **33**(5), 5583–5593 (2018)
15. J. Lai, X. Lu, X. Yu, W. Yao, J. Wen, S. Cheng, Distributed multi-DER cooperative control for master-slave-organized microgrid networks with limited communication bandwidth. IEEE Trans. Ind. Inf. **15**(6), 3443–3456 (2019)
16. N.M. Dehkordi, H.R. Baghaee, N. Sadati, J.M. Guerrero, Distributed noise-resilient secondary voltage and frequency control for islanded microgrids. IEEE Trans. Smart Grid **10**(4), 3780–3790 (2019)
17. J. Lai, X. Lu, Nonlinear mean-square power sharing control for ac microgrids under distributed event detection. IEEE Trans. Ind. Inf. **17**(1), 219–229 (2021)
18. S. Shrivastava, B. Subudhi, S. Das, Noise-resilient voltage and frequency synchronisation of an autonomous microgrid. IET Gener. Transm. Distrib. **13**(2), 189–200 (2018)
19. X. Wang, H. Zhang, C. Li, Distributed finite-time cooperative control of droop-controlled microgrids under switching topology. IET Renew. Power Gener. **11**(5), 707–714 (2017)
20. J. Lai, H. Zhou, X. Lu, X. Yu, W. Hu, Droop-based distributed cooperative control for microgrids with time-varying delays. IEEE Trans. Smart Grid **7**(4), 1775–1789 (Jul. 2016)
21. C. Dou, D. Yue, Z. Zhang, K. Ma, MAS-based distributed cooperative control for dc microgrid through switching topology communication network with time-varying delays. IEEE Syst. J. **13** (1), 615–624 (Jul. 2019)
22. J. Lai, X. Lu, X. Yu, A. Monti, H. Zhou, Distributed voltage regulation for cyber-physical microgrids with coupling delays and slow switching topologies. IEEE Trans. Syst., Man, Cybern. Syst. **50**(1), 100–110 (Jan. 2020)
23. A. Afshari, M. Karrari, H.R. Baghaee, G.B. Gharehpetian, S. Karrari, Cooperative fault-tolerant control of microgrids under switching communication topology. IEEE Trans. Smart Grid (2019). https://doi.org/10.1109/TSG.2019.2944768
24. D. Williams, *Probability with Martingales* (Cambridge University Press, Cambridge, 1991)
25. S. Liu, T. Li, L. Xie, M. Fu, J.F. Zhang, Continuous-time and sampled-data based average consensus with logarithmic quantizers. Automatica **49**(11), 3329–3336 (2013)
26. S. Lanzisera, D. Zats, K.S.J. Pister, Radio frequency time-of-flight distance measurement for low-cost wireless sensor localization. IEEE Sens. J. **11**(3), 837–845 (2011)
27. J. Lai, X. Lu, A. Monti, G.P. Liu, Stochastic distributed pinning control for co-multi-inverter networks with a virtual leader. IEEE Trans. Circuits Syst. II Exp. Briefs **67**(10), 2094–2098 (2020)
28. M. Rahmani-Andebili, Chapter 9: Cooperative distributed energy scheduling in microgrids, in *Electric Distribution Network Management and Control* (Springer, 2018), pp. 235–254
29. M. Rahmani-Andebili, Analyzing the effects of problem parameters on the operation cost of the networked microgrids, in *2020 IEEE Kansas Power and Energy Conference (KPEC)* (2020)

Chapter 3
Application of Optimization Techniques in the Design and Operation of Microgrid

Yixin Liu, Li Guo, Chengshan Wang, and Ruosong Hou

Abstract This chapter introduces a dynamic design framework of microgrid considering the future information of load growth, unit investment cost variation, and device degradation. The stochastic optimization and robust optimization techniques are utilized to deal with the long-term uncertainty of energy price and the short-term uncertainties of renewable energy-based distributed generation and load demands, respectively, among the design horizons. To speed up the design process and reduce the computational complexity, a typical scenario set generation method based on mixed integer multi-objective linear programming is applied. The microgrid design process is conducted in a dynamic perspective, which can provide not only the optimal capacities of distributed energy resources (DERs) but also the optimal investment timing for the stakeholders. Compared with the traditional microgrid design method where the DERs are installed in the first year, the proposed design framework shows better performance in terms of investment economy, renewable energy utilization rate, and fund recovery speed; therefore, it is more economically friendly to the stakeholders in the strategic planning design of microgrids.

Keywords Dynamic design of microgrid · Stochastic optimization · Robust optimization · Investment economy

3.1 Introduction

Microgrid planning and design is to determine the construction scheme satisfying the power demand, with comprehensive considerations of the load profile, distributed energy resource (DER) operating condition, and system status [1]. Different from the planning of utility power grid, the planning and design of microgrid is highly coupled with the operation optimization strategy due to its smaller scale and

Y. Liu (✉) · L. Guo · C. Wang · R. Hou
School of Electrical and Information Engineering, Tianjin University, Tianjin, China
e-mail: liuyixin@tju.edu.cn

M. Rahmani-Andebili (ed.), *Design, Control, and Operation of Microgrids in Smart Grids*, Power Systems, https://doi.org/10.1007/978-3-030-64631-8_3

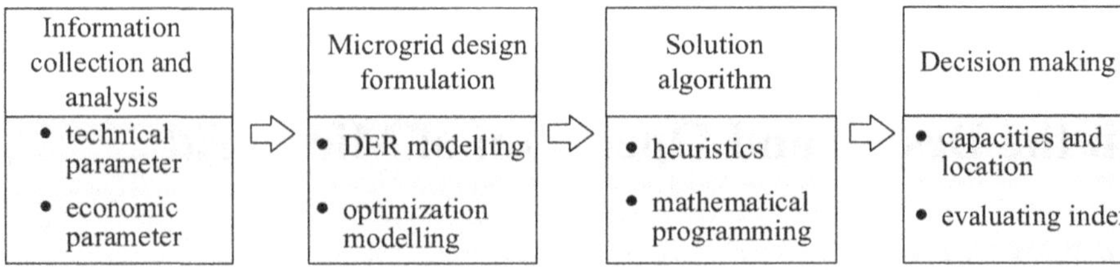

Fig. 3.1 Microgrid design process

sensitiveness to the randomness of renewable energy. As a result, microgrid design should comprehensively consider the influence of operation information on its economy, reliability, and environmental performance during the whole design horizon, making it more complex and requiring appropriate modelling and optimization techniques. Generally, the design process of microgrid can be demonstrated as Fig. 3.1.

3.1.1 Information Collection and Analysis

The primary premise of realizing the reasonable planning and design of microgrid is to collect and analyze the relevant data accurately, which can be divided into two categories: technical parameters and economic parameters.

Technical parameters mainly include historical load demand and growth trend, environmental data, historical distribution information of renewable energy such as wind speed and illumination, network structure information of existing system, and relevant physical parameters of alternative DERs. Economic parameters include macroeconomic parameters such as investment interest rate, inflation rate, and discount rate, as well as microeconomic parameters such as electricity price, fuel price, and device unit price, which will ultimately affect the economic evaluation performance of each scheme.

When analyzing the distribution characteristics of renewable energy and load demand, there are two methods according to whether the uncertainties are considered. Deterministic analysis method mainly refers to the historical data of resources and load demand of the microgrid. A typical application is to use 8760 hours of historical data to analyze the operation of microgrid [2]. This method is simple, but the results obtained in this way have certain limitations and cannot fully reflect the possible operation status of the system. Therefore, uncertainty analysis method that is able to capture the random characteristics of DER generation and load is critical to make the design scheme more reasonable and reliable.

3.1.2 Microgrid Design Formulation

3.1.2.1 Optimization Modelling

The optimization model of microgrid design can be divided into three parts: objective function, decision variables, and associated constraints. The design objective is up to the specific application scenarios and requirements of customer, including different aspects such as economic efficiency, system reliability, and environmental impacts. The objective function can be one or more of them, corresponding to single objective and multi-objective programming models [3]. The decision variables of microgrid design usually involve the type, capacity, and access location of DERs. Some researches have carried out the optimal design of microgrid structure according to the specific load demand. For example, ref. [4] and ref. [5] established a planning framework that considers not only the sizing of DERs but also the grid topology, i.e., AC, DC, or hybrid AC/DC of the stand-alone microgrids. Moreover, the investment timing needs to be considered when conducting the microgrid design in a dynamic perspective. The constraints mainly include two parts: constraints related to the planning layer, e.g., the installation capacity constraints, total cost constraints, and investment payback period constraints, and constraints related to the operating layer, such as power balance (electricity, heat, cold, and gas) constraints, power output constraints, ramp rate constraints, runtime constraints, and renewable energy penetration constraints.

3.1.2.2 Uncertainty Modelling

There exists a lot of uncertain information in the planning and operation stages of microgrid, which can be divided into five categories: load information, resource information, device information, market information, and policy information. It is necessary to consider the influence of these information when making the design scheme and operation scheme of microgrid.

1. **Uncertainty of load information**

Load forecasting can be divided into ultra short-term, short-term, medium-term, and long-term forecasting according to the time scale. The microgrid planning and design mainly focuses on the long-term load growth trend, and the daily operation simulation can be implemented via the typical load demand scenarios. The load demands, including electricity, heat, cold, and gas, are affected by multiple factors. Taking the electrical load as an example, different types of load (industrial, commercial, residential, etc.) have different consumption characteristics. Meanwhile, the meteorology also has an obvious impact on electrical load. The change of weather conditions such as temperature and cloudy and sunny weather will cause the variation of electrical load. Therefore, it is important to analyze the key factors that affect the load variation to improve the accuracy of load forecasting and take into account the uncertainty of load in the microgrid design process.

2. **Uncertainty of Resource**

The power generation of renewable energy-based DERs (RE-DERs) such as wind turbine (WT) and photovoltaic (PV) has obvious uncertainties [6]. The forecasting accuracy of renewable energy has a significant impact on the estimation of the output level of RE-DERs, which will further affect the performance of the design scheme and operation scheme of microgrid. In the operation stage, the ultra short-term and short-term forecasting of power generation of RE-DERs and load demand are utilized to optimize the dispatching scheme of microgrid. In the planning and design stage, the analysis of statistical distribution characteristics and stochastic characteristics of RE-DERs is required to evaluate the utilization of renewable energy resources.

3. **Uncertainty of Devices**

The normal operation of devices in microgrid will directly affect its reliability of power supply. For the grid-connected microgrid, it can still maintain power supply based on the support of external power grid in the majority of cases when internal emergency failure occurs. For the stand-alone microgrid, however, device failure is likely to cause system power failure and reduce the reliability of power supply. Consequently, it is of importance to evaluate the influence of device state uncertainty caused by fault on power supply reliability in the microgrid planning and design process.

4. **Uncertainty of Market**

For microgrid design, it is necessary to consider the influence of macroeconomic parameters such as loan interest rate and inflation rate, as well as microeconomic parameters such as electricity price, fuel price, and device unit investment price, and select appropriate uncertain objects and optimization techniques according to the research tendency [7]. In the microgrid operation stage, the uncertainty of market information, like the bidding price and clearing price, needs to be considered when making day-ahead operation plan and intra-day economic dispatch decision [8].

5. **Uncertainty of Policy**

In the early stage of the development of renewable energy, many countries and regions have introduced renewable energy generation subsidy policies to support the development of renewable energy industry due to its high cost for power generation. However, with the cost reduction, the proportion of renewable energy power generation in many countries has been greatly increased, and some countries began to reduce subsidies [9]. The change of subsidy policy has greatly impacted the economy of microgrid design. Moreover, the transaction mechanism between microgrid and external grid is still under development. The uncertainty of the release time and details of these policies will also affect the formulation of operation strategy and design scheme of microgrid.

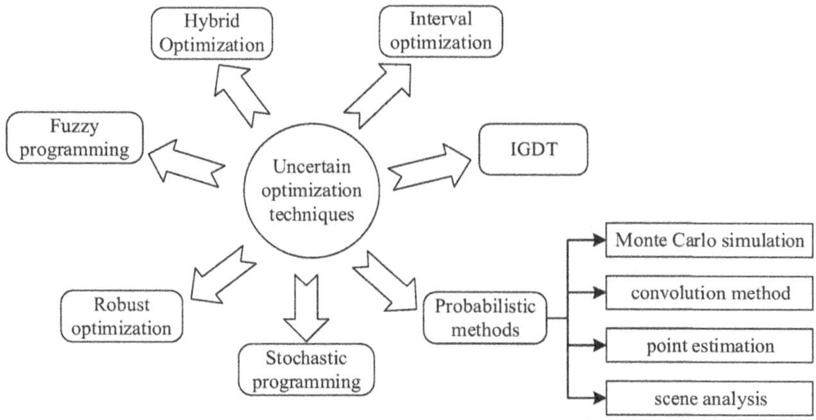

Fig. 3.2 Uncertainty optimization techniques

Aiming at the problem of microgrid design and operation with multiple uncertain information, a variety of optimization techniques can be used, which are summarized in Fig. 3.2.

- The probabilistic methods use random variables to describe the uncertain information and analyze the uncertain optimization problems based on the probability theory. The main probabilistic methods include Monte Carlo simulation method, convolution method, point estimation method, and scene analysis method [10].
- Although the stochastic programming method also uses random variables to describe the uncertain information, it seeks for a solution with expected minimum (or maximum) cost in an optimized way [11, 12].
- Robust optimization method applies uncertain sets to describe the possible realization of uncertain information and seeks to minimize (or maximize) the cost of the research problem in the "worst" scenario.
- Fuzzy programming uses fuzzy sets to describe uncertain information, and the evaluation given for the research problem is also fuzzy.
- The hybrid method takes advantage of different uncertainty optimization techniques according to the characteristics of uncertain information. For example, it usually combines probabilistic method with fuzzy programming to study uncertain information with randomness and fuzziness. In addition, the stochastic approach is combined with model predictive control (MPC) method to enhance the adaptive ability of optimization model to the uncertainties [13, 14].
- Interval optimization uses interval variables instead of point variables to describe the uncertain information and determines the upper and lower limits of the cost considering the uncertainty based on interval computing theory [15].
- Information gap decision theory (IGDT) can capture the uncertain information which cannot be described by random variables and fuzzy variables. The objective is to find the optimal solution under the maximum adverse effects of uncertain information.

3.1.3 Solution Algorithms

According to the formulated model, different solution algorithms can be utilized to solve the microgrid design problem, which can be divided into two major categories: heuristic algorithms and mathematical programming algorithms. Heuristic algorithms, including genetic algorithm [2], particle swarm optimization [16], tabu search algorithm [17], artificial bee colony algorithm [18], and differential evolution algorithm [19], are usually used to deal with NP-hard problems and obtain an acceptable approximate optimal solution. Mathematical programming algorithms are able to ensure the optimality of the solution but have strict requirements for the model form. The common mathematical programming algorithms include the nonlinear programming [20], mixed-integer linear programming (MILP) [21], and mixed-integer nonlinear programming [4, 5].

In most of the existing works, few researches consider both short-term and long-term uncertainties associated with planning decision and operation. Moreover, lots of the works conduct the microgrid design problem in a static perspective, where the unit investment cost and efficiency of devices remain unchanged among the design horizons. The design scheme is implemented in a way that all DERs are installed in the first year, which will result in redundant reserve capacity in the initial stage of the microgrid for the possible load growth in the future and cause unnecessary energy waste [22].

In this chapter, a dynamic microgrid design approach that considers both the long-term and short-term uncertainties, as well as the future information of load growth, unit investment cost variation, and device degradation, is introduced. The stochastic optimization and robust optimization techniques are applied to address the uncertainties, and the design process is formulated as an MILP model. The proposed design model can provide a cost-effective investment strategy for the stakeholders with optimal installation capacities of DERs and investment timing. Moreover, the conservative degree of the MILP model can be adjusted flexibly according to the risk aversion degree of stakeholders.

This chapter is organized as follows. In Sect. 3.2, the overall framework of the dynamic design approach is provided. Then a typical scenario set generation method and an MILP model for microgrid design are introduced. Section 3.3 elaborates the solution process of the model. In Sect. 3.4, numerous simulations are conducted and analyzed, and Sect. 3.5 gives the major conclusions.

3.2 Dynamic Design and Operation Model of Microgrid

3.2.1 Overall Framework

The overall framework of the proposed optimization model for the design and operation of microgrid is demonstrated in Fig. 3.3. The input data includes the

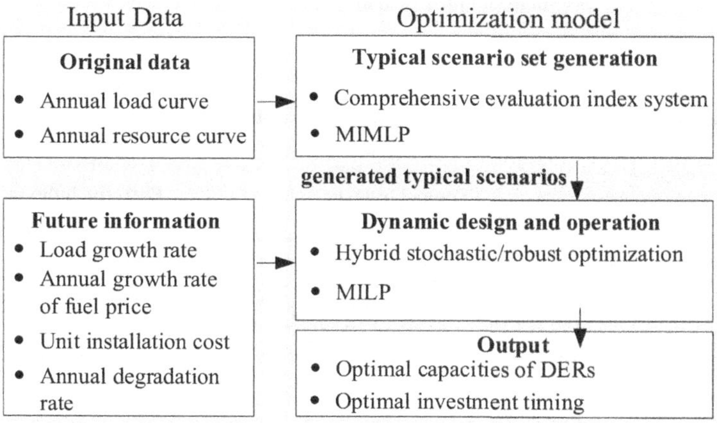

Fig. 3.3 Overall framework of the proposed optimization model for the design and operation of microgrid

annual load curve and resource curve, normally with hourly resolution, for the typical scenario set generation model and the future information for the dynamic design model. The typical scenario set generation model is formulated as a mixed-integer multi-objective linear programming (MIMLP) problem based on comprehensive evaluation indexes [23]. The purpose of the typical scenario set generation model is to provide the typical scenarios and the corresponding weights for the microgrid design and operation model to simplify the calculation complexity. The microgrid design model seeks for the optimal solution with excepted minimum cost, supporting the stakeholders to determine the optimal capacities of DERs and the corresponding investment timing.

3.2.2 *Typical Scenario Set Generation Model*

The key point of the typical scenario set generation model is to extract as much effective information as possible from the original data with limited scenarios. To this end, a multi-objective optimization model is established to comprehensively consider the performance of the typical scenarios in terms of scenario number and deviation with the original data.

3.2.2.1 Evaluation Indexes for Typical Scenario Set Generation

The key information from the original data includes the statistical information, time-series information, and extreme information of the load demand and resource, e.g.,

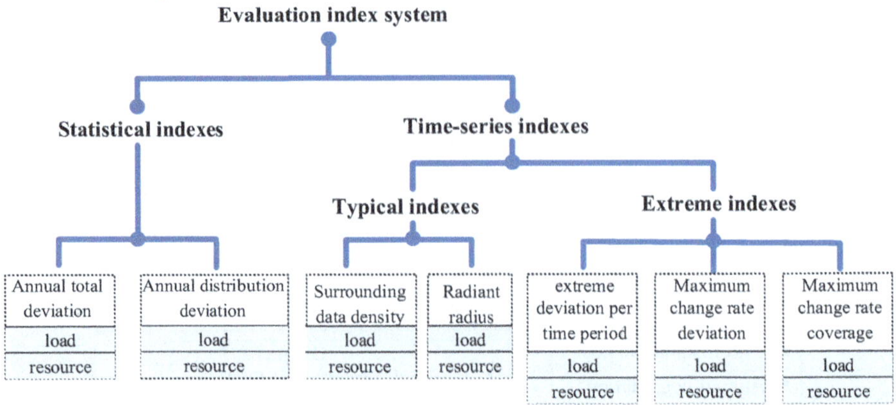

Fig. 3.4 Evaluation index system for the typical scenario set generation

irradiance and wind speed. Therefore, an evaluation index system shown in Fig. 3.4 is established.

1. **Statistical Indexes**

The statistical indexes are formulated as

$$\Delta C = \frac{\left|\sum_{d \in D} \omega_d \cdot C_d - C_{\text{year}}\right|}{C_{\text{year}}} \tag{3.1}$$

$$\Delta S = \frac{\left|\sum_{d \in D} \omega_d \cdot S_d - S_{\text{year}}\right|}{S_{\text{year}}} \tag{3.2}$$

$$\Delta P = \frac{1}{24}\sum_{t=1}^{24} \frac{\left|\sum_{d \in D} \omega_d P_{d,t}^{\text{typ}} - \sum_{d_0 \in D_0} P_{d_0,t}^{\text{ori}}\right|}{\sum_{d_0 \in D_0} P_{d_0,t}^{\text{ori}}} \tag{3.3}$$

$$\Delta W = \frac{1}{24}\sum_{t=1}^{24} \frac{\left|\sum_{d \in D} \omega_d W_{d,t}^{\text{typ}} - \sum_{d_0 \in D_0} W_{d_0,t}^{\text{ori}}\right|}{\sum_{d_0 \in D_0} W_{d_0,t}^{\text{ori}}} \tag{3.4}$$

where ΔC and ΔS denote the annual total load deviation and resource deviation, respectively, and ΔP and ΔW are the annual load demand distribution deviation and resource distribution deviation, respectively. D and D_0 indicate the set of typical scenarios and the original data; d and d_0 are the indexes for date; ω_d represents the weight coefficient of the typical scenario d; C_d and S_d denote the total amount of load demand and resource of typical scenario d; C_{year} and S_{year} denote the annual total amount of load and resource of the original data; $P_{d,t}^{\text{typ}}$, $W_{d,t}^{\text{typ}}$ and $P_{d_0,t}^{\text{ori}}$, $W_{d_0,t}^{\text{ori}}$ indicate the load power and resource value of typical scenario d and original data at time period t, respectively.

In (3.1)–(3.4), ω_d is equal to 0 if the corresponding day is not selected as the solution of typical scenarios. Therefore, the statistical indexes are used to evaluate the performance of the typical scenario set in terms of total amount deviation and distribution deviation of load demand and resource. (3.1) and (3.2) embody the relative errors of the total load and resources between typical scenarios and those of the original data. (3.3) and (3.4) indicate the average of the load power deviations and resource deviations of the whole time periods.

2. **Time-Series Indexes**

In addition to the total amount information and distribution information, the typical scenario set generation model also needs to balance the performance of typicality and extreme scenario information extraction. To ensure the typicality and representativeness, the following two indexes are considered:

$$\rho_i = \sum_{j \in I_S, j \neq i} \chi\left(d_{ij} - d_c\right)$$

$$\chi(x) = \begin{cases} 1, & x < 0 \\ 0, & x \geq 0 \end{cases} \tag{3.5}$$

$$I_s = \{1, 2, \ldots, \text{card}(D_0)\}$$

$$\delta = \begin{cases} \min\limits_{j \in I_s^i} \{d_{ij}\}, I_s^i \neq O \\ \max\limits_{j \in I_s^i} \{d_{ij}\}, I_s^i = O \end{cases} \tag{3.6}$$

$$I_s^i = \{k \in I_s : \rho_k > \rho_i\}$$

where ρ_i represents the surrounding data density of scenario i, d_{ij} denotes the Euclidean distance between the data vectors of scenario i and scenario j, d_c is the truncation distance, I_s is the set of alternatives for typical scenarios, and *card*() indicates the number of elements in the set. Each element in I_s represents the sequence number of the alternatives; δ expresses the radiant radius of scenario i; O means empty set; I_s^i is the set of scenarios whose surrounding data density is larger than that of scenario i.

The surrounding data density of scenario i shows the number of data points within the truncation distance. The radiant radius of scenario i is defined as the distance between itself and the global farthest point if the scenario i is the point with the global maximum data density. Otherwise, it is defined as the distance between itself and the nearest data point whose density is larger than itself.

From the above definition, it can be seen that the scenarios with larger surrounding data density and radiation radius are more likely to be selected as the typical scenarios. However, the original data may contain some extreme scenarios that have great influence on the operation strategy and design scheme of microgrid. These extreme scenarios are usually isolated with the other scenarios; thus, they may be ignored by the conventional typical scenario generation algorithm. To this end, other eight indexes are considered:

$$\Delta L_{\max,t} = \left| \frac{\max_{d \in D} P_{d,t}^{\text{typ}} - \max_{d_0 \in D_0} P_{d_0,t}^{\text{ori}}}{\max_{d_0 \in D_0} P_{d_0,t}^{\text{ori}}} \right| \tag{3.7}$$

$$\Delta L_{\min,t} = \left| \frac{\min_{d \in D} P_{d,t}^{\text{typ}} - \min_{d_0 \in D_0} P_{d_0,t}^{\text{ori}}}{\min_{d_0 \in D_0} P_{d_0,t}^{\text{ori}}} \right| \tag{3.8}$$

$$\Delta S_{\max,t} = \left| \frac{\max_{d \in D} W_{d,t}^{\text{typ}} - \max_{d_0 \in D_0} W_{d_0,t}^{\text{ori}}}{\max_{d_0 \in D_0} W_{d_0,t}^{\text{ori}}} \right| \tag{3.9}$$

$$\Delta S_{\min,t} = \left| \frac{\min_{d \in D} W_{d,t}^{\text{typ}} - \min_{d_0 \in D_0} W_{d_0,t}^{\text{ori}}}{\min_{d_0 \in D_0} W_{d_0,t}^{\text{ori}}} \right| \tag{3.10}$$

$$\Delta L^{\text{ra}} = \max_t \left| \frac{\max_{d \in D} \left| P_{d,t+1}^{\text{typ}} - P_{d,t}^{\text{typ}} \right| - \max_{d_0 \in D_0} \left| P_{d_0,t+1}^{\text{ori}} - P_{d_0,t}^{\text{ori}} \right|}{\max_{d_0 \in D_0} \left| P_{d_0,t+1}^{\text{ori}} - P_{d_0,t}^{\text{ori}} \right|} \right| \tag{3.11}$$

$$\Delta S^{\text{ra}} = \max_t \left| \frac{\max_{d \in D} \left| W_{d,t+1}^{\text{typ}} - W_{d,t}^{\text{typ}} \right| - \max_{d_0 \in D_0} \left| W_{d_0,t+1}^{\text{ori}} - W_{d_0,t}^{\text{ori}} \right|}{\max_{d_0 \in D_0} \left| W_{d_0,t+1}^{\text{ori}} - W_{d_0,t}^{\text{ori}} \right|} \right| \tag{3.12}$$

$$\Delta L^{\text{cov}} = \frac{\text{card}\left(\begin{array}{l} \max_{d \in D,\, t} \left| P_{d,t+1}^{\text{typ}} - P_{d,t}^{\text{typ}} \right| \geq \\ \left\{ \left| P_{d_0,t+1}^{\text{ori}} - P_{d_0,t}^{\text{ori}} \right| \middle| \forall d_0 \in D_0, \forall t \right\} \end{array} \right)}{\text{card}(D_0)}, \ t = (1, 2, \cdots T-1) \tag{3.13}$$

$$\Delta S^{\text{cov}} = \frac{\text{card}\left(\begin{array}{l} \max_{d \in D,\, t} \left| W_{d,t+1}^{\text{typ}} - W_{d,t}^{\text{typ}} \right| \geq \\ \left\{ \left| W_{d_0,t+1}^{\text{ori}} - W_{d_0,t}^{\text{ori}} \right| \middle| \forall d_0 \in D_0, \forall t \right\} \end{array} \right)}{\text{card}(D_0)}, \ t = (1, 2, \cdots, T-1) \tag{3.14}$$

where $\Delta L_{\max,\,t}/\Delta L_{\min,\,t}$ and $\Delta S_{\max,\,t}/\Delta S_{\min,\,t}$ denote the maximum/minimum load deviations and resource deviations at time period t, respectively; ΔL^{ra} and ΔS^{ra}

represent the maximum load demand/resource change rate deviation; ΔL^{cov} and ΔS^{cov} indicate the load demand/resource change rate coverage.

In (3.7)–(3.10), the four indexes are applied to capture the extremum information such as heavy load or poor resource situations of the original data. The other four indexes formulated as (3.11)–(3.14) are able to reflect the time-series fluctuation information of load demand and resources between adjacent periods. The load demand /resource change rate coverage represents the relative position of the maximum fluctuation in the variation value of the original data, which is equal to 1 if the day with maximum fluctuation scenario of load/resources in adjacent periods is involved into the typical scenario set.

3.2.2.2 MIMLP Formulation for Typical Scenario Set Generation

Considering the comprehensive performance of the typical scenario set in terms of the proposed indexes and the scenario number, an MIMLP model is established as

$$\begin{cases} \min z_1 = \sum u_i \\ \min z_2 = \sum\limits_m \dfrac{\|\boldsymbol{H}_m\boldsymbol{\omega} - \boldsymbol{f}_m\|_1}{\|\boldsymbol{f}_m\|_1} \\ \max\ z_3 = \boldsymbol{\rho\omega} \\ \max z_4 = \boldsymbol{\delta\omega} \end{cases}$$
$$\boldsymbol{H} = [\boldsymbol{H}_1, \boldsymbol{H}_2, \cdots \boldsymbol{H}_m]^{\mathrm{T}}, \boldsymbol{f} = [\boldsymbol{f}_1, \boldsymbol{f}_2, \cdots \boldsymbol{f}_m]^{\mathrm{T}} \tag{3.15}$$
$$\boldsymbol{H}_m = \begin{bmatrix} data_{1,1} & \cdots & data_{N,1} \\ \vdots & \ddots & \vdots \\ data_{1,24} & \cdots & data_{N,24} \end{bmatrix}, \boldsymbol{f}_m = \begin{bmatrix} \sum\limits_i data_{i,1} \\ \vdots \\ \sum\limits_i data_{i,24} \end{bmatrix}$$

subject to

$$\begin{cases} (a)\ \omega_i \le u_i \cdot N, \forall i \in [1,N] \mid i \in \mathrm{Z}^+ \\ (b) \sum \omega_i = N, \forall i \in [1,N] \mid i \in \mathrm{Z}^+ \\ (c) \text{-} \boldsymbol{f} \cdot \alpha \le \boldsymbol{H\omega} - \boldsymbol{f} \le \boldsymbol{f} \cdot \alpha, \alpha \in R - R^- \\ (d)\ u_{k_1} + u_{k_2} + \cdots + u_{k_j} \ge \beta, k_j \in \left\{ \begin{array}{l} \Delta L_{\max,t} \le \varepsilon_1 \cup \Delta L_{\min,t} \le \varepsilon_2 \cup \\ \Delta S_{\max,t} \le \varepsilon_3 \cup \Delta S_{\min,t} \le \varepsilon_4 \cup \\ \Delta L^{ra} \le \varepsilon_5 \cup \Delta S^{ra} \le \varepsilon_6 \cup \\ \Delta L_t^{\mathrm{cov}} \ge \varepsilon_7 \cup \Delta S_t^{ra} \ge \varepsilon_8 \end{array} \right\} \\ (e)\ \omega_i \in R - R^-, \forall i \in [1,N] \mid i \in \mathrm{Z}^+ \\ (f)\ u_i \in \{0,1\} \end{cases} \quad (3.16)$$

In (3.15), the first objective function minimizes the number of typical scenarios. u_i is equal to 1 if scenario i is involved into the typical scenario set; otherwise, it is equal to 0. The second objective function optimizes the performance of typical scenarios with regard to statistical indexes of (3.1)–(3.4). m denotes the number of type of the original data. For example, if the original data includes load demand, wind speed, and solar irradiance, m is equal to 3, and the element "data" in $\boldsymbol{H}_m$ and $\boldsymbol{f}_m$ satisfies data $\in$ {load, wind, solar}. Each element in matrix $\boldsymbol{H}$ denotes the load or resource value for each day. N is the total number of original days. Each element in $\boldsymbol{f}$ represents the total load or resource amount of the original data at the same time period. $\boldsymbol{\omega}$ is the weight coefficient column vector of typical scenarios. The third and fourth objective functions maximize the performance of typical scenarios in terms of indexes of (3.5) and (3.6). $\boldsymbol{\rho} = [\rho_1, \rho_2, \cdots, \rho_N]$ and $\boldsymbol{\delta} = [\delta_1, \delta_2, \cdots, \delta_N]$ are the row vectors of surrounding data density and radiation radius of typical scenarios, respectively.

In (3.16), constraint (a) guarantees that the weight coefficient ω_i is equal to 0 if the corresponding u_i is equal to 0; constraint (b) ensures the sum of the weights of all typical scenarios is equal to N; constraint (c) limits the total load and resource deviation at each time period within a specified range, α is a predefined coefficient; constraint (d) provides a way to choose the number of extreme scenarios being considered in the solution by adjusting the value of β, $\beta \le z_1$. ε_1~ε_8 are predefined extreme index constraints; constraint (e) means that the weight coefficient ω_i is a nonnegative real number; and constraint (f) illustrates that the u_i is a binary variable.

The decision variables of the MIMLP model include the weight coefficients ω_i and u_i. Instead of determining the scenario number by the users themselves, the MIMLP model is able to find an optimal solution of ω_i and u_i with limited scenario numbers.

3.2.3 Dynamic Planning and Design Model

3.2.3.1 Stochastic Programming Model

The major concern for the stakeholders of microgrid lies in the investment economy of the project, which is essentially an economic planning and design problem that optimizes both the long-term DER investment decision and short-term operation dispatch of DERs. In addition, considering the high cost of microgrid construction, conducting business loans from banks to relieve the financial pressure at the initial stage of the project is the mainstream investment form of microgrid. Therefore, in the proposed microgrid design model, the investment cost minimization over the design horizon is considered as the objective function with a set of constraints related to investment decisions and operation. Without loss of generality, a dynamic planning and design model for a stand-alone microgrid that consists of diesel generator (DG), PV, and battery energy storage system (BESS) is introduced in this chapter, which is formulated as

$$
\begin{aligned}
&\min \sum_{n=1}^{N_{\mathrm{Y}}} \left(C_{inv,n} + C_{opr,n}\right) \\
&= \sum_{k\in\{\mathrm{DG},\mathrm{PV}\}} \left(X_1^k P_{\min}^k e_1^k (1 - r_{\mathrm{loan}}) + \sum_{n=2}^{N_{\mathrm{Y}}} \frac{X_n^k P_{\min}^k e_n^k}{(1+r)^{n-1}} - \frac{\sum_{n=1}^{N_{\mathrm{Y}}} X_n^k P_{\min}^k e_{\mathrm{sal}}^k}{(1+r)^{N_{\mathrm{Y}}\text{-}1}} \right) + \\
&\sum_{k\in\{\mathrm{BESS}\}} \left(X_1^k P_{\min}^k e_1^k (1 - r_{\mathrm{loan}}) + \sum_{n=2}^{N_{\mathrm{Y}}} \frac{X_n^k P_{\min}^k e_n^k}{(1+r)^{n-1}} - \frac{S_{N_{\mathrm{Y}}}^k e_{\mathrm{sal}}^k}{(1+r)^{N_{\mathrm{Y}}\text{-}1}} \right) \\
&+ \sum_{k} \sum_{n=1}^{N_{\mathrm{loan}}} \left(\frac{X_1^k P_{\min}^k e_1^k r_{\mathrm{loan}} \mathrm{APR}(1+\mathrm{APR})^{N_{\mathrm{loan}}}}{(1+\mathrm{APR})^{N_{\mathrm{loan}}} - 1} \frac{1}{(1+r)^{n-1}} \right) + \\
&\sum_{k\in\{\mathrm{DG},\mathrm{PV}\}} \sum_{n=1}^{N_{\mathrm{Y}}} \frac{\sum_{i=1}^{n} X_i^k P_{\min}^k e_{\mathrm{om}}^k}{(1+r)^{n-1}} + \sum_{k\in\{\mathrm{BESS}\}} \sum_{n=1}^{N_{\mathrm{Y}}} \frac{S_n^k e_{\mathrm{om}}^k}{(1+r)^{n-1}} + \\
&\sum_{n=1}^{N_{\mathrm{Y}}} \sum_{d\in D, s\in S, t\in T} \omega_{n,d} \cdot \omega_s \cdot \left(\frac{P_{n,d,s,t}^{\mathrm{DG}} \pi_{\mathrm{fuel},n,s} e_{\mathrm{DG}}}{(1+r)^{n-1}} + \frac{P_{n,d,s,t}^{\mathrm{loss}} q_{\mathrm{L}}}{(1+r)^{n-1}} \right)
\end{aligned}
\tag{3.17}
$$

subject to

$$
\mu P_{\min}^{\mathrm{DG}} \le P_{n,d,s,t}^{\mathrm{DG}} \le \left(\sum_{i=1}^{n} X_i^{\mathrm{DG}} \right) \times P_{\min}^{\mathrm{DG}} \quad \forall n, \forall d, \forall s, \forall t \tag{3.18}
$$

$$\underline{\mathrm{SOC}} \cdot S_n^{\mathrm{BESS}} \leq S_{n,d,s,t}^{\mathrm{BESS}} \leq \overline{\mathrm{SOC}} \cdot S_n^{\mathrm{BESS}} \quad \forall n, \forall d, \forall s, \forall t \tag{3.19}$$

$$0 \leq P_{\mathrm{ch},n,d,s,t}^{\mathrm{BESS}} \leq \gamma_{\mathrm{ch}} \cdot S_n^{\mathrm{BESS}} \quad \forall n, \forall d, \forall s, \forall t \tag{3.20}$$

$$0 \leq P_{\mathrm{dis},n,d,s,t}^{\mathrm{BESS}} \leq \gamma_{\mathrm{dis}} \cdot S_n^{\mathrm{BESS}} \quad \forall n, \forall d, \forall s, \forall t \tag{3.21}$$

$$S_n^{\mathrm{BESS}} = \left(1 - B_n^{\mathrm{BESS}}\right) S_{n-1}^{\mathrm{BESS}} + X_n^{\mathrm{BESS}} P_{\min}^{\mathrm{BESS}} \quad \forall n \tag{3.22}$$

$$\begin{aligned} & S_{n,d,s,t}^{\mathrm{BESS}} - S_{n,d,s,t-1}^{\mathrm{BESS}} - P_{\mathrm{ch},n,d,s,t}^{\mathrm{BESS}} \eta_{\mathrm{ch}}^{\mathrm{BESS}} + P_{\mathrm{dis},n,d,s,t}^{\mathrm{BESS}} / \eta_{\mathrm{dis}}^{\mathrm{BESS}} = 0 \\ & \forall n, \forall d, \forall s, \forall t \end{aligned} \tag{3.23}$$

$$\sum_t P_{\mathrm{ch},n,d,s,t}^{\mathrm{BESS}} \eta_{\mathrm{ch}}^{\mathrm{BESS}} - \sum_t P_{\mathrm{dis},n,d,s,t}^{\mathrm{BESS}} / \eta_{\mathrm{dis}}^{\mathrm{BESS}} = 0 \quad \forall n, \forall d, \forall s, \forall t \tag{3.24}$$

$$\begin{aligned} R_n = {} & E^{\max} \cdot X_n^{\mathrm{BESS}} P_{\min}^{\mathrm{BESS}} - \sum_{d \in D,\, s \in S,\, t \in T} \omega_{n,d} \cdot \omega_s P_{\mathrm{dis},n,d,s,t}^{\mathrm{BESS}} \\ & + \left(1 - B_n^{\mathrm{BESS}}\right) R_{n-1} \quad \forall n \end{aligned} \tag{3.25}$$

$$R_n \geq 0 \tag{3.26}$$

$$B_n^{\mathrm{BESS}} X_n^{\mathrm{BESS}} P_{\min}^{\mathrm{BESS}} - \xi X_n^{\mathrm{BESS}} P_{\min}^{\mathrm{BESS}} \geq 0 \quad \forall n, \xi \in (0, 1) \tag{3.27}$$

$$\widehat{P}_{n,d,t}^{\mathrm{PV}} = \sum_{i=1}^{n} X_i^{\mathrm{PV}} P_{\min}^{\mathrm{PV}} (1 - \sigma_{\mathrm{PV}})^{n-i} p_{i,d,t}^{\mathrm{PV}} \quad \forall n, \forall d, \forall t \tag{3.28}$$

$$0 \leq X_n^k P_{\min}^k \leq \overline{S_n^k} \quad \forall n, \forall k \tag{3.29}$$

$$0 \leq P_{n,d,s,t}^{\mathrm{aban}} \leq \overline{P_{n,d,s,t}^{\mathrm{aban}}} \quad \forall n, \forall d, \forall s, \forall t \tag{3.30}$$

$$0 \leq P_{n,d,s,t}^{\mathrm{loss}} \leq \overline{P_{n,d,s,t}^{\mathrm{loss}}} \quad \forall n, \forall d, \forall s, \forall t \tag{3.31}$$

$$\begin{aligned} & P_{n,d,s,t}^{\mathrm{DG}} + P_{n,d,t}^{\mathrm{PV}} + P_{\mathrm{dis},n,d,s,t}^{\mathrm{BESS}} \eta_{\mathrm{dis}}^{\mathrm{BESS}} - P_{\mathrm{ch},n,d,s,t}^{\mathrm{BESS}} / \eta_{\mathrm{ch}}^{\mathrm{BESS}} \\ & + P_{n,d,s,t}^{\mathrm{loss}} - P_{n,d,s,t}^{\mathrm{aban}} = P_{n,d,t}^{\mathrm{load}} \quad \forall n, \forall d, \forall s, \forall t \end{aligned} \tag{3.32}$$

In (3.17), n, d, s, and t are the indexes for the year, typical day, scenario, and time period; S and T are the sets for scenarios and dispatching cycle. Each scenario s represents a realization of fuel price in nth year; N_{Y} is the design horizon; $C_{inv,\,n}$, $C_{opr,\,n}$ denote the total cost related to investment and operation. The second and third lines illustrate the investment cost of DG, PV, and BESS. X_n^k indicates the installed number of minimum unit of DER k in nth year. $P_{\min}^k$ is the capacity of minimum installed unit of DER k, and e_n^k is the corresponding unit cost in nth year. In the dynamic design model, the loan behavior is considered only in the first year of the project; thus, the first term of second and third lines means the actual expenditure in the first year. r_{loan} is the loan ratio. The rest item of second and third lines of (3.17) represents the net present value (NPV) of investment cost and the salvage value of DG, PV, and BESS. r denotes the discount rate. The salvage value of BESS is calculated in a different way with DG and PV as the replacement of BESS should be considered among the design horizons. e_{sal}^{k} is the unit salvage value of DER k. S_{n}^{BESS} indicates the efficient capacity of BESS in nth year. The rest of the parts of (3.17) represents the operational cost, including the annual loan interest formulated as the fourth line, the operation and maintenance (O&M) cost of DER k shown as the fifth line, and the fuel consumption cost and load loss cost for all possible scenarios demonstrated as the sixth line of (3.17). N_{loan} denotes the loan cycle; APR represents the annual interest rate; e_{om}^{k} is the unit O&M cost of DER k; $\omega_{n,\,d}$ and ω_s indicate the weight coefficients of typical day d in nth year and the scenario s, respectively; $P_{n,d,s,t}^{\text{DG}}$ and $P_{n,d,s,t}^{\text{loss}}$ express the output power of DG and unmet load power; $\pi_{\text{fuel},\,n,\,s}$ represents the fuel price; e_{DG} denotes the fuel consumption per kWh of DG; q_{L} means the penalty cost for load loss.

Constraint (3.18) limits the output power of DG, and μ denotes the minimum load rate of DG. Constraints (3.19)–(3.27) give the operational constraints of BESS. Constraints (3.19)–(3.21) ensure the remaining capacity $S_{n,d,s,t}^{\text{BESS}}$, charge power $P_{\text{ch},n,d,s,t}^{\text{BESS}}$, and discharge power $P_{\text{dis},n,d,s,t}^{\text{BESS}}$ of BESS within a reasonable range. γ_{ch} and γ_{dis} denote the maximum charge and discharge rate of BESS, respectively. Constraint (3.22) demonstrates the capacity of BESS in nth year, where the binary variable B_{n}^{BESS} is equal to 1 if the BESS needs to be replaced in nth year, otherwise, it is equal to 0. The relationship of the remaining capacity of adjacent time periods is described by constraint (3.23). $\eta_{\text{ch}}^{\text{BESS}}$ and $\eta_{\text{dis}}^{\text{BESS}}$ are the charge and discharge efficiency, respectively. Constraint (3.24) guarantees that the remaining capacity at the last time period is equal to the initial capacity. Constraints (3.25) and (3.26) formulate the life model of BESS. The available discharge capacity of BESS in nth year is represented by (3.25). $E^{\max}$ indicates the maximum discharge electricity of BESS per unit capacity. Constraint (3.27) ensures that new BESS will be installed only when it needs to be replaced. In (3.28), $\widehat{P}_{n,d,t}^{\text{PV}}$ means the forecast PV generation, and $p_{i,d,t}^{\text{PV}}$ denotes the output power of PV per unit capacity. The annual degradation rate of PV panel is considered in the dynamic design model and expressed by σ_{PV}. Constraint (3.29) limits the upper bound of installation capacity of DER k in nth year. Constraints (3.30) and (3.30) demonstrate the maximum allowed values for abandoned power and load curtailment. Constraint (3.31) ensures the power balance of microgrid.

In the above formulated model, constraints (3.22) and (3.25)–(3.27) are nonlinear with the product of binary variable and continuous variable and are linearized by introducing auxiliary variables L_n', L_n'', L_n''' as

$$S_{n}^{\text{BESS}} = S_{n-1}^{\text{BESS}} - L_n' + X_{n}^{\text{BESS}} P_{\min}^{\text{BESS}} \tag{3.33}$$

$$\begin{aligned} R_n = {} & E^{\max} X_{n}^{\text{BESS}} P_{\min}^{\text{BESS}} - \sum_{d\in D,\, s\in S,\, t\in T} \omega_{n,d} \cdot \omega_s P_{\text{dis},n,d,s,t}^{\text{BESS}} \\ & + R_{n-1} - L_n'' \end{aligned} \tag{3.34}$$

$$
\begin{aligned}
& E^{\max} X_n^{\mathrm{BESS}} P_{\min}^{\mathrm{BESS}} - \sum_{d \in D, s \in S, t \in T} \omega_{n,d} \cdot \omega_s P_{\mathrm{dis},n,d,s,t}^{\mathrm{BESS}} \\
& + R_{n-1} - L_n'' \geq 0
\end{aligned} \tag{3.35}
$$

$$
L_n''' P_{\min}^{\mathrm{BESS}} - \tau X_n^{\mathrm{BESS}} P_{\min}^{\mathrm{BESS}} \geq 0 \tag{3.36}
$$

The auxiliary variables should meet the following constraints to guarantee equivalence:

$$
\begin{cases}
0 \leq L_n' \leq B_n^{\mathrm{BESS}} \cdot \overline{S_n^{\mathrm{BESS}}} \\
S_{n-1}^{\mathrm{BESS}} - \left(1 - B_n^{\mathrm{BESS}}\right) \cdot \overline{S_n^{\mathrm{BESS}}} \leq L_n' \leq S_{n-1}^{\mathrm{BESS}}
\end{cases} \tag{3.37}
$$

$$
\begin{cases}
0 \leq L_n'' \leq B_n^{\mathrm{BESS}} \cdot \overline{R_n} \\
R_{n-1} - \left(1 - B_n^{\mathrm{BESS}}\right) \cdot \overline{R_n} \leq L_n'' \leq R_{n-1}
\end{cases} \tag{3.38}
$$

$$
\begin{cases}
0 \leq L_n''' \leq B_n^{\mathrm{BESS}} \cdot \overline{S_n^{\mathrm{BESS}}} \\
X_n^{\mathrm{BESS}} P_{\min}^{\mathrm{BESS}} - \left(1 - B_n^{\mathrm{BESS}}\right) \cdot \overline{S_n^{\mathrm{BESS}}} \leq L_n''' \leq X_n^{\mathrm{BESS}} P_{\min}^{\mathrm{BESS}}
\end{cases} \tag{3.39}
$$

$$
L_n' = B_n^{\mathrm{BESS}} S_{n-1}^{\mathrm{BESS}}, L_n'' = B_n^{\mathrm{BESS}} R_{n-1}, L_n''' = B_n^{\mathrm{BESS}} X_n^{\mathrm{BESS}} P_{\min}^{\mathrm{BESS}} \tag{3.40}
$$

After the above linearization, the dynamic design model becomes an MILP form. The decision variables can be summarized as

$$
\begin{cases}
\boldsymbol{x} = \left(X_n^k, B_n^{BESS}, S_n^{\mathrm{BESS}}, R_n, L_n', L_n'', L_n'''\right)^{\mathrm{T}} \quad \forall k, \forall n \\
\boldsymbol{y} = \begin{pmatrix} P_{n,d,s,t}^{\mathrm{DG}}, P_{\mathrm{dis},n,d,s,t}^{\mathrm{BESS}}, P_{\mathrm{ch},n,d,s,t}^{\mathrm{BESS}}, S_{n,d,s,t}^{\mathrm{BESS}}, \\ P_{n,d,s,t}^{\mathrm{loss}}, P_{n,d,s,t}^{\mathrm{aban}}, P_{n,d,t}^{\mathrm{PV}}, P_{n,d,t}^{\mathrm{load}} \end{pmatrix}^{\mathrm{T}} \\
\forall n, \forall d, \forall s, \forall t \\
X_n^k \in \mathbb{Z}_+, B_n^{BESS} \in \{0, 1\}, \boldsymbol{y} \in 0 \cup \mathbb{R}_+
\end{cases} \tag{3.41}
$$

where $\boldsymbol{x}$ and $\boldsymbol{y}$ are vectors related to planning design and operational decisions, respectively.

As mentioned above, the fuel price in nth year is considered as a stochastic variable and described by a stochastic process of geometric Brownian motion (GBM) as

$$
\pi_{\mathrm{fuel},n} = \pi_{\mathrm{fuel},n-1} \times e^{r_{\mathrm{fuel}}} + \varepsilon \tag{3.42}
$$

where ε is a random component that is normally distributed with a mean of zero and a standard deviation calculated by using the first difference of past 25-year prices from 1996 to 2020 [24].

For the convenience of expression, the compact form of the dynamic design model is formulated as

$$
\begin{aligned}
&\min_{x,y} \left\{ \boldsymbol{c}^{\mathrm{T}} \boldsymbol{x} + \boldsymbol{d}^{\mathrm{T}} \boldsymbol{y} \right\} \\
&s.t. \quad \boldsymbol{Ax} \geq \boldsymbol{a}, \\
&\qquad \boldsymbol{Bx} = \boldsymbol{b}, \\
&\qquad \boldsymbol{Qy} \geq \boldsymbol{q}, \\
&\qquad \boldsymbol{Gy} = \boldsymbol{g}, \\
&\qquad \boldsymbol{Wy} \geq \boldsymbol{h} - \boldsymbol{Fx}, \\
&\qquad \boldsymbol{My} = \boldsymbol{m} - \boldsymbol{Rx}, \\
&\qquad \boldsymbol{Iy} = \widehat{\boldsymbol{u}}, \\
&\qquad \forall n, \forall d, \forall s, \forall t
\end{aligned}
\tag{3.43}
$$

where $\boldsymbol{c}$ and $\boldsymbol{d}$ are the coefficient column vectors of objective function; $\boldsymbol{A}$, $\boldsymbol{B}$, $\boldsymbol{Q}$, $\boldsymbol{G}$, $\boldsymbol{W}$, $\boldsymbol{F}$, $\boldsymbol{M}$, $\boldsymbol{R}$, and $\boldsymbol{I}$ are the coefficient matrixes of variables corresponding to the constraints; and $\boldsymbol{a}$, $\boldsymbol{b}$, $\boldsymbol{q}$, $\boldsymbol{g}$, $\boldsymbol{h}$, and $\boldsymbol{m}$ are constant column vectors.

The eighth line in (3.43) denotes that the uncertainties of PV power and load demand are ignored in the stochastic programming model, and the values of PV power and load demand are equal to the forecast power, which is shown as

$$
\widehat{\boldsymbol{u}} = \left[\widehat{P}_{n,d,t}^{\mathrm{PV}}, \widehat{P}_{n,d,t}^{\mathrm{load}} \right]^{\mathrm{T}} \quad \forall n, \forall d, \forall t \tag{3.44}
$$

where $\widehat{P}_{n,d,t}^{\mathrm{load}}$ is the forecast value of load demand.

3.2.3.2 Hybrid Stochastic/Robust Optimization Model

To better counter unexpected consequences of PV generation and load demand, robust optimization technique is applied, and the stochastic programming model is extended to a hybrid stochastic/robust optimization model. In the integrated framework, the fluctuation of PV generation and load demand is considered within the following set $\boldsymbol{U}$:

$$
\boldsymbol{U} := \left\{ \begin{array}{l}
\boldsymbol{u} = \left[u_{n,d,t}^{\mathrm{PV}}, u_{n,d,t}^{\mathrm{load}} \right]^{\mathrm{T}}, \quad \forall n, \forall d, \forall t \mid \\
u_{n,d,t}^{\mathrm{PV}} \in \left[\widehat{P}_{n,d,t}^{\mathrm{PV}} - \Delta P_{n,d,t}^{\mathrm{PV}}, \widehat{P}_{n,d,t}^{\mathrm{PV}} + \Delta P_{n,d,t}^{\mathrm{PV}} \right]; \\
u_{n,d,t}^{\mathrm{load}} \in \left[\widehat{P}_{n,d,t}^{\mathrm{load}} - \Delta P_{n,d,t}^{\mathrm{load}}, \widehat{P}_{n,d,t}^{\mathrm{load}} + \Delta P_{n,d,t}^{\mathrm{load}} \right].
\end{array} \right\} \tag{3.45}
$$

where $u_{n,d,t}^{\mathrm{PV}}, u_{n,d,t}^{\mathrm{load}}$ are the uncertain variables related to PV generation and load demand and $\Delta P_{n,d,t}^{\mathrm{PV}}$ and $\Delta P_{n,d,t}^{\mathrm{load}}$ are the maximum forecast deviations.

The hybrid stochastic/robust optimization model is then established as

$$
\begin{aligned}
&\min\left\{\boldsymbol{c}^{\mathrm{T}}\boldsymbol{x}+\max_{\boldsymbol{u}\in U}\min_{\boldsymbol{y}\in\Omega(\boldsymbol{x},\boldsymbol{u})}\boldsymbol{d}^{\mathrm{T}}\boldsymbol{y}\right\}\\
&s.t.\quad \boldsymbol{Ax}\geq\boldsymbol{a},\\
&\qquad\ \ \boldsymbol{Bx}=\boldsymbol{b},\\
&\qquad\ \ \boldsymbol{Qy}\geq\boldsymbol{q},\\
&\qquad\ \ \boldsymbol{Gy}=\boldsymbol{g},\\
&\qquad\ \ \boldsymbol{Wy}\geq\boldsymbol{h}-\boldsymbol{Fx},\\
&\qquad\ \ \boldsymbol{My}=\boldsymbol{m}-\boldsymbol{Rx},\\
&\qquad\ \ \boldsymbol{Iy}=\boldsymbol{u},\\
&\qquad\ \ \forall n,\forall d,\forall s,\forall t
\end{aligned}
\tag{3.46}
$$

In (3.46), the inner layer max-min optimization problem is able to find the "worst scenario" of $\boldsymbol{u}$ that may lead to a deterioration of operating costs. $\Omega(\boldsymbol{x},\boldsymbol{u})$ indicates the feasible region of $\boldsymbol{y}$ with fixed $\boldsymbol{x}$ and $\boldsymbol{u}$. The outer layer optimization problem is to realize an optimal solution with expected minimum cost under the realization of uncertainties of PV generation and load demand.

3.3 Solution Approach

The typical scenario set generation model formulated as (3.15) and (3.16) is an MIMLP problem and can be solved by a two-stage fuzzy programming method [25–27]. The following is the specific solution process:

1. Obtain the maximum and minimum value of each objective in (3.15) under the single objective scenario separately.
2. Construct the membership functions of each objective according to the respective optimization direction based on the maximum and minimum value of each objective, and seek the optimal membership degree that is no larger than the membership functions of each objective, and obtain the solution via the MILP method.
3. Maximize the average value of all the membership functions of each objective in the second stage and achieve the solution, which is used to test the solution of the first step. If the solution of the second step is the same as that of the first step, the two-stage fuzzy algorithm is valid and the solution is unique. Otherwise, the solution of the second stage is taken as the suboptimal solution.
4. Output the results of the typical scenario set generation model.

The microgrid design model formulated as (3.46) is essentially an MILP problem with appropriate transformations. The Benders decomposition method is used to decompose the model into a master problem (MP) and a subproblem (SP). Firstly, the SP is formulated as

$$\begin{aligned}
&\max_{u\in U}\ \min_{y\in\Omega(x,u)}\ \boldsymbol{d}^{\mathrm{T}}\boldsymbol{y} \\
&s.t.\quad \boldsymbol{Qy}\geq\boldsymbol{q}, && \leftarrow\boldsymbol{\gamma} \\
&\qquad\ \boldsymbol{Gy}=\boldsymbol{g}, && \leftarrow\boldsymbol{\tau} \\
&\qquad\ \boldsymbol{Wy}\geq\boldsymbol{h}-\boldsymbol{Fx}, && \leftarrow\boldsymbol{\psi} \\
&\qquad\ \boldsymbol{My}=\boldsymbol{m}-\boldsymbol{Rx}, && \leftarrow\boldsymbol{\theta} \\
&\qquad\ \boldsymbol{Iy}=\boldsymbol{u} && \leftarrow\boldsymbol{\varpi}
\end{aligned} \tag{3.47}$$

where $\boldsymbol{\gamma}, \boldsymbol{\tau}, \boldsymbol{\psi}, \boldsymbol{\theta}, \boldsymbol{\varpi}$ denote the dual variable vectors of the corresponding constraints.

Define $\boldsymbol{y}^{\mathrm{DUAL}}=[\boldsymbol{\gamma}^{\mathrm{T}},\boldsymbol{\tau}^{\mathrm{T}},\boldsymbol{\psi}^{\mathrm{T}},\boldsymbol{\theta}^{\mathrm{T}},\boldsymbol{\varpi}^{\mathrm{T}}]^{\mathrm{T}}$. The SP can be converted into the following maximization problem based on the strong duality theory:

$$\begin{aligned}
&\max_{\boldsymbol{u},\boldsymbol{y}^{\mathrm{DUAL}}}\left(\boldsymbol{q}^{\mathrm{T}}\boldsymbol{\gamma}+\boldsymbol{g}^{\mathrm{T}}\boldsymbol{\tau}+(\boldsymbol{h}-\boldsymbol{Fx})^{\mathrm{T}}\boldsymbol{\psi}+(\boldsymbol{m}-\boldsymbol{Rx})^{\mathrm{T}}\boldsymbol{\theta}+\boldsymbol{u}^{\mathrm{T}}\boldsymbol{\varpi}\right) \\
&s.t.\quad \boldsymbol{Q}^{\mathrm{T}}\boldsymbol{\gamma}+\boldsymbol{G}^{\mathrm{T}}\boldsymbol{\tau}+\boldsymbol{W}^{\mathrm{T}}\boldsymbol{\psi}+\boldsymbol{M}^{\mathrm{T}}\boldsymbol{\theta}+\boldsymbol{I}^{\mathrm{T}}\boldsymbol{\varpi}\leq\boldsymbol{d}, \\
&\qquad\ \boldsymbol{\gamma}\geq 0,\boldsymbol{\psi}\geq 0, \\
&\qquad\ \boldsymbol{u}\in\boldsymbol{U}
\end{aligned} \tag{3.48}$$

Note that the problem in (3.48) is a bilinear optimization problem. The optimal solution is achieved only when the realization of $\boldsymbol{u}$ is an extreme point of the uncertainty set $\boldsymbol{U}$ [28]. Then, the uncertainty set $\boldsymbol{U}$ can be rewritten as

$$\boldsymbol{U}:=\left\{\begin{array}{l}
\boldsymbol{u}=\left[u_{n,d,t}^{\mathrm{PV}},u_{n,d,t}^{\mathrm{load}}\right]^{\mathrm{T}}\ | \\
u_{n,d,t}^{\mathrm{PV}}=\widehat{P}_{n,d,t}^{\mathrm{PV}}-B_{n,d,t}^{\mathrm{PV}}\Delta P_{n,d,t}^{\mathrm{PV}}, \\
\sum\limits_{t\in T}B_{n,d,t}^{\mathrm{PV}}\leq\Gamma_{\mathrm{PV}},\quad \forall n,d; \\
u_{n,d,t}^{\mathrm{load}}=\widehat{P}_{n,d,t}^{\mathrm{load}}+B_{n,d,t}^{\mathrm{load}}\Delta P_{n,d,t}^{\mathrm{load}} \\
\sum\limits_{t\in T}B_{n,d,t}^{\mathrm{load}}\leq\Gamma_{\mathrm{load}},\forall n,d.
\end{array}\right\} \tag{3.49}$$

where $B_{n,d,t}^{\mathrm{PV}}$ and $B_{n,d,t}^{\mathrm{load}}$ are binary variables related to the realization of worst scenario, whose value is equal to 1 if the corresponding uncertain variable is taken as the boundary of $\boldsymbol{U}$ and equal to 0 otherwise. $\boldsymbol{\Gamma}^{\mathrm{PV}}$ and $\boldsymbol{\Gamma}^{\mathrm{load}}$ represent the uncertain budgets used to adjust the conservatism of the optimal solutions. Both $\boldsymbol{\Gamma}^{\mathrm{PV}}$ and $\boldsymbol{\Gamma}^{\mathrm{load}}$ are integers, denoting the maximum number of periods when $\boldsymbol{u}$ is taken as the boundary of $\boldsymbol{U}$ in the decision-making process.

Then, the SP can be recast into the following MILP form by linearization techniques, which is expressed as

$$\begin{aligned}
&\max_{u,y^{\mathrm{DUAL}}} \left(q^{\mathrm{T}}\gamma + g^{\mathrm{T}}\tau + (h - Fx)^{\mathrm{T}}\psi + (m - Rx)^{\mathrm{T}}\theta + \widehat{u}^{\mathrm{T}}\varpi + \Delta u^{\mathrm{T}}B' \right) \\
&s.t. \qquad Q^{\mathrm{T}}\gamma + G^{\mathrm{T}}\tau + W^{\mathrm{T}}\psi + M^{\mathrm{T}}\theta + I^{\mathrm{T}}\varpi \le d, \\
&\qquad\qquad 0 \le B' \le \mathrm{M}B, \\
&\qquad\qquad \varpi - \mathrm{M}(1 - B) \le B' \le \varpi, \\
&\qquad\qquad \sum_{t\in T} B^{\mathrm{PV}}_{n,d,t} \le \Gamma_{\mathrm{PV}}, \qquad \forall n, d \\
&\qquad\qquad \sum_{t\in T} B^{\mathrm{load}}_{n,d,t} \le \Gamma_{\mathrm{load}}, \qquad \forall n, d \\
&\qquad\qquad \gamma \ge 0, \psi \ge 0, \\
&\qquad\qquad u \in U
\end{aligned} \tag{3.50}$$

where $\Delta u = \left[-\Delta P^{\mathrm{PV}}_{n,d,t}, \Delta P^{\mathrm{load}}_{n,d,t}\right]^{\mathrm{T}}$, $B = \left[B^{\mathrm{PV}}_{n,d,t}, B^{\mathrm{load}}_{n,d,t}\right]^{\mathrm{T}}$, and $B' = \left[B'^{\mathrm{PV}}_{n,d,t}, B'^{\mathrm{load}}_{n,d,t}\right]^{\mathrm{T}}$ is the auxiliary variable vectors for linearization and M is a real number larger enough.

For the MP, it includes not only the constraints of SP but also the cutting plane constructed by the optimal solution of SP. The form of MP is formulated as

$$\begin{aligned}
&\min_{x,\eta,y} \left\{ c^{\mathrm{T}}x + \eta \right\} \\
&s.t. \quad Ax \ge a, \\
&\qquad Bx = b, \\
&\qquad \eta \ge d^{\mathrm{T}}y, \\
&\qquad Qy \ge q, \\
&\qquad Gy = g, \\
&\qquad Wy \ge h - Fx, \\
&\qquad My = m - Rx, \\
&\qquad Iy = u, \\
&\qquad \forall n, d, s, t
\end{aligned} \tag{3.51}$$

After the above transformation, the hybrid optimization model is recast into the MP (3.51) and SP (3.50). The solving procedure is as follows:

1. Define a realization of u in U as the initial worst scenario. Set the upper bound of the objective function of (3.46) UB $= +\infty$ and the lower bound LB $= -\infty$. Fix the iteration counter $it = 1$.
2. Solve the MP based on the worst scenario u^*_{it}, and achieve the optimal solution $\left(x^*_{it}, \eta^*_{it}, y^{1*}, \cdots, y^{it*}\right)^{\mathrm{T}}$ and update LB$=c^{\mathrm{T}}x^*_{it} + \eta^*_{it}$.
3. Solve the SP based on x^*_{it} and obtain the objective function value of (3.50) and the worst scenario, denoted as $f^*_{it}\left(x^*_{it}\right)$ and u^*_{it+1}. Update $UB = \min\left(UB, c^{\mathrm{T}}x^*_{it} + f^*_{it}\left(x^*_{it}\right)\right)$.

4. If $UB - LB \leq \varsigma$, ς is a given gap tolerance, return $\boldsymbol{x}_{it}^{*}, \boldsymbol{y}^{it*}$and terminate. Otherwise, add a new cutting plane constraint into the MP. Let $it = it + 1$ and go to step 2.

3.4 Case Study

To verify the effectiveness and advantages of the microgrid design model, numerous case studies are conducted based on practical load and resource data in Northwest China. The MIMLP model for typical scenario set generation and MILP model for microgrid dynamic design are solved by IBM ILOG CPLEX 12.6.3 with MATLAB R2013a.

Table 3.1 summarizes the main economic and operational parameters for the microgrid design problem. The unit costs of PV and BESS are compiled based on refs. [29, 30] and are presented in Fig. 3.5.

Figure 3.6 illustrates the optimal solution of typical scenario set generation model with the selected typical days as well as their weight coefficients. Due to the seasonality of solar radiation and electricity load, the typical scenario is selected by quarter. Taking the first quarter as an example, the 68th day and 56th day are allocated with larger weights. This is because the data density and the radiation

Table 3.1 Economic and operational parameters for MILP

Parameters	Value
Design horizon (year)	15
Discount rate	5%
Annual degradation rate of PV panel	1%
Loan cycle (year)	10
Loan rate	6.5%
Maximum charge/discharge rate of BESS	1C
Maximum/minimum SOC	0.05/0.95
Charge/discharge efficiency of BESS	0.95
Penalty cost for load loss (¥/kWh)	10
Fuel consumption per kWh of DG (kWh/L)	0.33
Minimum load rate of DG	0.30
Maximum discharge electricity of BESS per unit capacity (kWh)	2000
Minimum installed unit of PV (kW)	1
Minimum installed unit of BESS (kWh)	1
Minimum installed unit of DG (kW)	500
Initial unit installation cost of PV (¥/kW)	4000
Initial unit installation cost of BESS (¥/kWh)	2000
Initial unit installation cost of DG (¥/kW)	500
Unit O&M cost of PV (¥/kW per year)	40
Unit O&M cost of BESS (¥/kWh per year)	25
Unit O&M cost of DG (¥/kW per year)	50
Salvage value rate	0.03

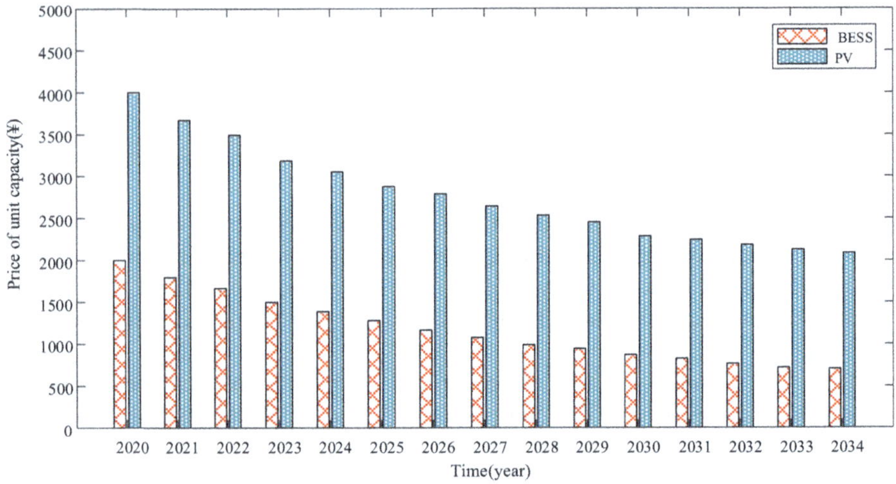

Fig. 3.5 Unit costs of PV and BESS among the design horizons

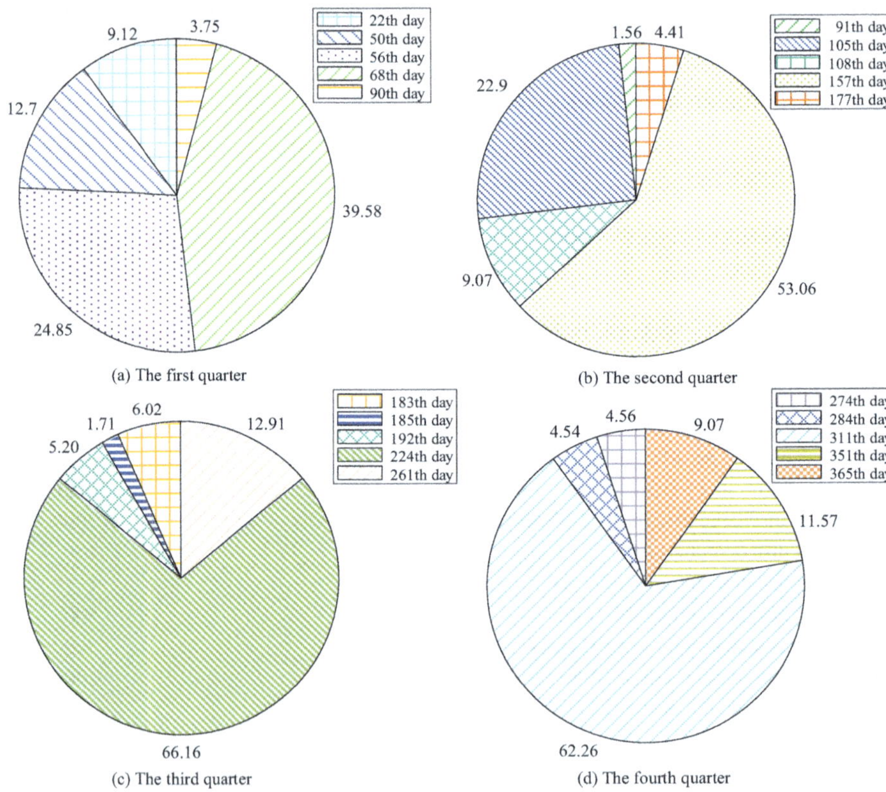

Fig. 3.6 The optimal solution of typical scenario set generation model

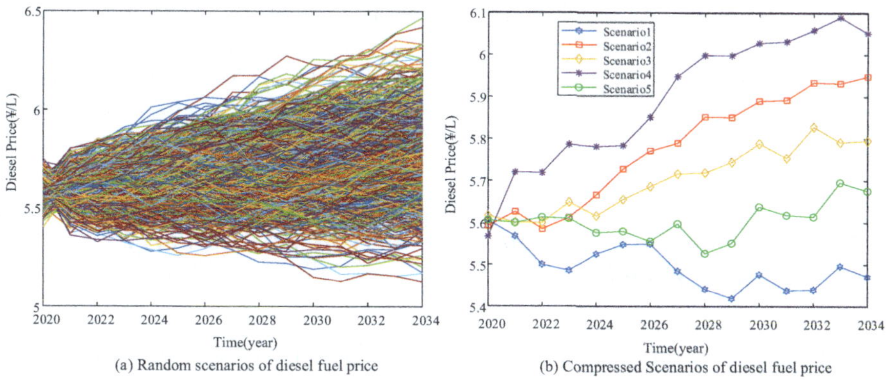

Fig. 3.7 The scenarios of diesel fuel price

Fig. 3.8 The sum of square error in different clustering number

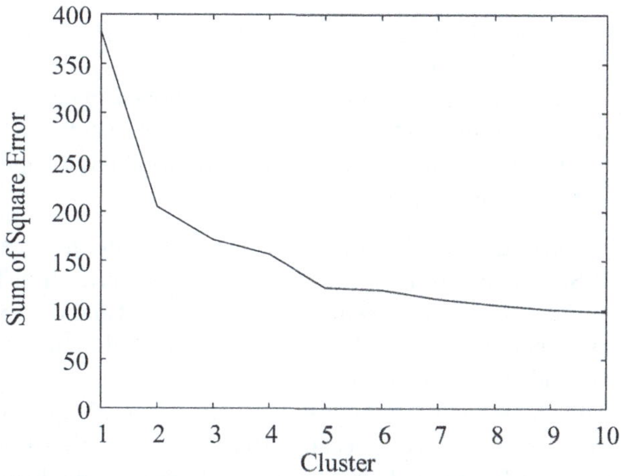

radius of these two days are obviously larger than the other days in the first quarter, denoting that they have strong representativeness. For some scenarios, like the 185th day in the third quarter, the radiation radius is large, but the data density is relatively small. These scenarios may be the points that represent extreme scenarios; thus, they are also involved in the solution to take into account the extreme information of the original data but with small weight coefficients.

To address the random characteristics of fuel price, 1000 scenarios are generated with the expected annual growth rate of diesel fuel price as 0.27% [31]. Then the scenarios are reduced to five scenarios based on k-means clustering method. Figure 3.7 demonstrates the generated random scenarios of diesel fuel price (Fig. 3.7a) as well as the compressed scenarios (Fig. 3.7b). The sum of square error [32] is used to determine the optimal number of clustering, and the value with the change of clustering number is shown in Fig. 3.8. It can be seen that the sum of square error reduces along with the increase of cluster number. However, the descending speed becomes very slow once the clustering number is greater than 5. Therefore, considering the accuracy and computational efficiency, five scenarios are preferred. The weight coefficients of each scenario are 0.144, 0.23, 0.259, 0.094, and 0.273, respectively.

3.4.1 Economic Performance of the Dynamic Microgrid Design Model

To verify the advantages of the dynamic microgrid design model, three cases under different investment strategies are conducted.

Case 1: The investment decision is implemented in a static perspective, where the future dynamic information such as load growth, fuel price variation, unit costs of PV and BESS, and the degradation of PV panel are ignored.
Case 2: Similar to case 1, the investment behavior occurs in the first year. However, the future dynamic information mentioned in case 1 is considered in case 2.
Case 3: The investment decision is made by the dynamic design model proposed in this chapter, in which not only the installation capacities of DERs but also the investment timing is considered.

Although the decisions are developed in the first year in case 1 and case 2, the BESS will also be replaced at the end of its service life. The loan ratio is 80% in case 1–case 3. The load growth rate is 3% in case 2 and case 3. Table 3.2 demonstrates the investment scheme of these three cases with the minimum installation units of DG, PV, and BESS in each year.

It can be seen that the installed capacity of PV in case 2 is larger than that of case 1 and case 3. This is because large PV generation is more economic to meet the needs for future load growth in case 2, whereas the load growth is not considered in case 1. For case 3, the PV investment decision is carried out in many years to reduce the initial cost and utilize the benefit of cost reduction of PV. Meanwhile, the PV is invested mainly in 2020 and 2027 in case 3, corresponding to the years with newly installed BESS. In 2027, the BESS is replaced for all three cases. The capacity of BESS in 2027 is the same with that in 2020 in case 1, as the PV generation and load demand are unchanged over the whole design horizon in case 1. For case 2 and case 3, considering the unit cost reduction of BESS shown in Fig. 3.5, more BESS investments are scheduled in 2027. The number of DG is the same in all three cases, which requires two DGs with a rated power of 500 kW in the first year.

Table 3.3 presents the yearly economic performance of the investment schemes shown in Table 3.2. The abandoned energy ratio denotes the ratio of total abandoned PV generation to the total PV generation in the corresponding year. The PV penetration rate and abandoned energy ratio in case 1 remain unchanged, as the dynamic information is ignored. For case 2 and case 3, the PV penetration rate decreases along with the load growth during 2020–2026 and 2027–2034 and has a

Table 3.2 Investment scheme of case 1-case 3

	Case								
	Case 1			Case 2			Case 3		
	Installed capacity			Installed capacity			Installed capacity		
Year	DG	PV	BESS	DG	PV	BESS	DG	PV	BESS
2020	2	3047	6238	2	4289	5718	2	2726	5303
2021	0	0	0	0	0	0	0	214	0
2022	0	0	0	0	0	0	0	50	0
2023	0	0	0	0	0	0	0	52	0
2024	0	0	0	0	0	0	0	55	0
2025	0	0	0	0	0	0	0	338	0
2026	0	0	0	0	0	0	0	0	0
2027	0	0	6238	0	0	8746	0	1232	9908
2028	0	0	0	0	0	0	0	140	0
2029	0	0	0	0	0	0	0	56	0
2030	0	0	0	0	0	0	0	56	0
2030	0	0	0	0	0	0	0	0	0
2032	0	0	0	0	0	0	0	0	0
2033	0	0	0	0	0	0	0	0	0
2034	0	0	0	0	0	0	0	0	0

significant improvement in 2027 due to the newly installed BESS. It can be seen that the abandoned energy ratio in case 2 is obviously higher than that of case 3 during 2020–2026, denoting that too much reserve capacity of PV is considered in case 2 in the initial stage of the project and causes unnecessary energy waste. Moreover, this will also bring greater financial pressure in the initial stage of the project for the stakeholder. For case 3, thanks to the dynamic investment strategy, the abandoned energy ratio keeps at a low level, denoting a better matching between power generation and consumption in each year and a higher utilization of PV generation.

The total costs, operating costs, and the total PV penetration rate and abandoned energy ratio are illustrated in Fig. 3.9. Although the total cost and operating cost of case 1 are the smallest, the design scheme is unable to cope with future load growth. It can be seen that the total cost as well as the operating cost in case 3 is lower than that in case 2, indicating the economic superiority of the proposed investment strategy. The total cost is reduced from ¥ 93,153,262 to ¥ 90,303,634 (reduced by 3.60%), and the operating cost is reduced from ¥ 62,223,128 to ¥ 59,699,885 (reduced by 4.06%). The total abandoned energy ratio over the design horizon is 17.26% in case 3 but reaches 22.69% in case 2. The total PV penetration rate is 68.56% in case 2 and 69.74% in case 3, which is similar.

The income of the microgrid lies in selling the electricity to the loads, and the selling price is set as ¥ 1.2/kWh. Table 3.3 also gives the present value of the cumulative net cash flow (C-NCF) of the three cases. It can be seen that the present value of C-NCF becomes positive in 2024 in case 3, but is postponed to 2026 in case 2. This denotes that the dynamic design strategy shows better performance in terms

Table 3.3 Yearly economic performance of case 1–case 3

Year	Item								
	PV penetration rate			Abandoned energy ratio			Present value of cumulative net cash flow($\times 10^4$ ¥)		
	Case 1	Case 2	Case 3	Case 1	Case 2	Case 3	Case 1	Case 2	Case 3
2020	70.62%	70.86%	66.35%	15.67%	39.89%	11.44%	−367.63	−489.32	−310.24
2021	70.62%	69.92%	66.50%	15.67%	38.29%	14.44%	−238.45	−396.94	−247.89
2022	70.62%	69.01%	65.43%	15.67%	36.63%	13.92%	−115.41	−304.23	−131.08
2023	70.62%	68.09%	64.27%	15.67%	34.95%	13.57%	1.76	−214.23	−20.99
2024	70.62%	66.97%	63.14%	15.67%	33.44%	13.28%	113.36	−129.97	80.58
2025	70.62%	65.80%	62.75%	15.67%	31.96%	19.56%	219.64	−50.59	118.91
2026	70.62%	64.64%	61.57%	15.67%	30.46%	17.87%	320.86	23.04	224.94
2027	70.62%	73.79%	75.92%	15.67%	17.40%	23.77%	−61.52	−493.46	−539.88
2028	70.62%	72.82%	75.71%	15.67%	15.20%	23.34%	30.29	−342.48	−336.44
2029	70.62%	71.76%	75.25%	15.67%	13.06%	21.71%	117.72	−198.03	−121.88
2030	70.62%	70.32%	74.78%	15.67%	11.36%	20.04%	372.92	135.98	240.54
2030	70.62%	68.89%	73.94%	15.67%	9.65%	17.74%	615.96	453.25	598.41
2032	70.62%	67.47%	72.85%	15.67%	7.94%	15.69%	847.42	751.94	939.99
2033	70.62%	65.81%	71.49%	15.67%	6.58%	13.91%	1067.87	1029.57	1262.66
2034	70.62%	63.85%	70.11%	15.67%	5.70%	12.16%	1178.70	1309.96	1594.92

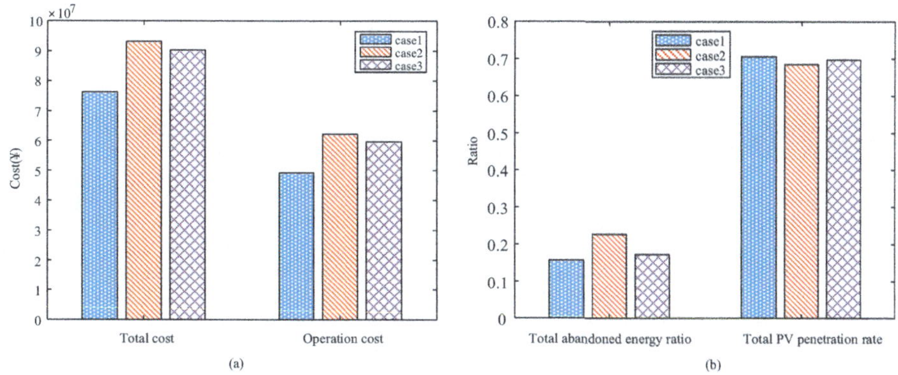

Fig. 3.9 Economic performance over the design horizon

of fund recovery speed. The present value of C-NCF in case 2 and case 3 becomes negative in 2027 due to investment of PV and BESS, and the final C-NCF in case 3 is the largest.

3.4.2 Sensitivity Analyses of the Dynamic Microgrid Design Model

3.4.2.1 Sensitivity Analysis for Load Growth Rate

The load growth rate has a great influence on the investment strategy and the associated economy of the microgrid design. Therefore, four cases are considered to evaluate the performance of the proposed dynamic microgrid design model.

Case 4: The load growth rate increases to 5%, and the investment strategy is the same as case 2.

Case 5: The load growth rate increases to 5%, and the investment strategy is the same as case 3.

Case 6: The load growth rate increases to 7%, and the investment strategy is the same as case 2.

Case 7: The load growth rate increases to 7%, and the investment strategy is the same as case 3.

Table 3.4 demonstrates the investment scheme of case 4–case 7 in each year. Comparing case 4 with case 6, it can be found that the initial installed capacities of DERs increase significantly with the increase of load growth rate when the investment strategy of case 2 is adopted. However, the load growth rate has little influence on the initial installed capacities of DERs when the dynamic investment strategy of case 3 is adopted. Due to faster load growth rate, a new DG is required in 2030 in case 7, ahead of case 5. When comparing case 4 and case 6 with case 5 and case 7, it

Table 3.4 Investment scheme of case 4–case 7

	Case											
	Case 4			Case 5			Case 6			Case 7		
	Installed capacity			Installed capacity			Installed capacity			Installed capacity		
Year	DG	PV	BESS	DG	PV	BESS	DG	PV	BESS	DG	PV	BESS
2020	3	5306	5823	2	2739	5375	4	6422	6108	2	2777	5580
2021	0	0	0	0	253	0	0	0	0	0	304	0
2022	0	0	0	0	85	0	0	0	0	0	121	0
2023	0	0	0	0	92	0	0	0	0	0	132	0
2024	0	0	0	0	96	0	0	0	0	0	142	0
2025	0	0	0	0	620	0	0	0	0	0	903	0
2026	0	0	0	0	0	0	0	0	0	0	0	0
2027	0	0	10,448	0	1535	12,113	0	0	12,247	0	1903	14,997
2028	0	0	0	0	299	0	0	0	0	0	458	0
2029	0	0	0	0	164	0	0	0	0	0	414	0
2030	0	0	0	0	117	0	0	0	0	1	197	0
2030	0	0	0	0	124	0	0	0	0	0	214	0
2032	0	0	0	1	0	0	0	0	0	0	232	0
2033	0	0	0	0	0	0	0	0	0	0	0	0
2034	0	0	0	0	0	0	0	0	0	0	0	0

can be seen that the initial installed capacities of DERs are significantly reduced under the dynamic investment strategy; thus, it is beneficial to reduce the initial investment cost.

The yearly economic performance of case 4–case 7 is shown in Table 3.5. Comparing case 2 with case 4 and case 6, it can be found that the PV penetration rate slightly increases with the increase of load growth rate. However, when compared case 3 with case 5 and case 7, it shows that the load growth rate has little influence on the PV penetration rate under the dynamic design model. In addition, the abandoned energy ratio increases significantly with the increase of load growth rate under the investment strategy of case 2, which can be easily found when case 2 is compared with case 4 and case 6. However, there is only a slight increase in abandoned energy ratio with the change of load growth rate under the proposed dynamic design strategy, showing a better adaptability to the load variation. The present value of C-NCF in Table 3.5 also shows better performance of the dynamic design strategy in terms of fund recovery speed, and the advantage of the dynamic design strategy is more obvious with the increase of load growth rate.

3.4.2.2 Sensitivity Analysis for Loan Ratio

In the proposed dynamic microgrid design model, the loan is considered to reduce the financial pressure at the initial stage of the project. The loan ratio has a significant impact on the present value of C-NCF. Therefore, the following three cases are considered to analyze the influence of the loan ratio:

Case 8: Loan ratio is set as 0%, denoting that no loan behavior occurs.
Case 9: Loan ratio is set as 40%.
Case 10: Loan ratio is set as 80%.

All three cases adopt the dynamic design model proposed in this chapter. The load growth rate is set as 3%, the same with case 3.

Figure 3.10 provides the present value of C-NCF of case 8–case 10. It can be seen that the final present value of C-NCF in 2034 of case 8 is the highest, indicating that the final profit of case 8 is the highest. However, the initial investment is much higher than that of case 9 and case 10. This will bring huge financial pressure to the stakeholder. For instance, the present value of C-NCF becomes positive until 2029 in case 8. For case 9 and case 10, it advances to 2026 and 2024, respectively. On the other hand, larger loan ratio means higher financial leverage, and the stakeholders will face greater risk on loan interest in the case of cash flow disruption. Therefore, an appropriate loan ratio needs to be determined to balance the financial pressure and risk.

Table 3.5 Yearly economic performance of case 4–case 7

Year	Item											
	PV penetration rate				Abandoned energy ratio				Present value of cumulative net cash flow ($\times 10^4$ ¥)			
	Case 4	Case 5	Case 6	Case 7	Case 4	Case 5	Case 6	Case 7	Case 4	Case 5	Case 6	Case 7
2020	72.12%	66.64%	73.29%	67.28%	50.55%	11.48%	58.48%	11.85%	−620.92	−314.08	−778.57	−325.13
2021	70.93%	66.63%	72.13%	67.02%	48.42%	14.15%	55.83%	14.55%	−561.03	−260.36	−757.90	−282.51
2022	69.69%	64.96%	70.66%	65.12%	46.25%	13.72%	53.24%	13.71%	−493.79	−146.92	−720.66	−170.22
2023	68.38%	63.41%	69.11%	63.18%	44.06%	13.34%	50.57%	13.21%	−423.49	−37.72	−672.09	−59.17
2024	66.76%	61.99%	67.11%	61.38%	42.07%	12.89%	48.12%	12.71%	−355.31	65.98	−619.91	48.21
2025	65.19%	61.79%	65.18%	61.25%	40.01%	23.20%	45.54%	25.99%	−288.08	60.64	−563.56	1.95
2026	63.71%	60.20%	63.35%	59.36%	37.82%	20.63%	42.80%	22.48%	−222.56	187.58	−504.61	152.97
2027	75.82%	77.34%	77.14%	78.77%	21.51%	23.80%	24.71%	23.83%	−829.89	−733.43	−1211.87	−981.01
2028	74.90%	77.16%	76.16%	78.48%	17.77%	23.80%	19.66%	23.82%	−630.09	−478.81	−962.79	−666.03
2029	73.37%	76.62%	74.24%	78.06%	14.56%	22.12%	15.35%	23.10%	−434.20	−196.37	−712.48	−323.65
2030	71.31%	75.95%	71.54%	77.21%	11.93%	19.83%	11.84%	20.13%	−18.12	248.75	−205.93	206.09
2031	69.22%	74.94%	68.64%	75.85%	9.34%	17.93%	8.58%	17.77%	378.49	685.90	276.39	748.35
2032	66.91%	73.11%	65.10%	73.89%	7.04%	15.08%	6.29%	16.19%	750.49	1106.38	721.01	1276.08
2033	63.99%	71.13%	61.31%	71.39%	5.72%	12.38%	4.61%	12.49%	1089.66	1518.63	1120.59	1807.71
2034	61.13%	68.94%	57.57%	68.15%	4.47%	9.92%	3.20%	9.71%	1428.36	1941.66	1512.41	2347.95

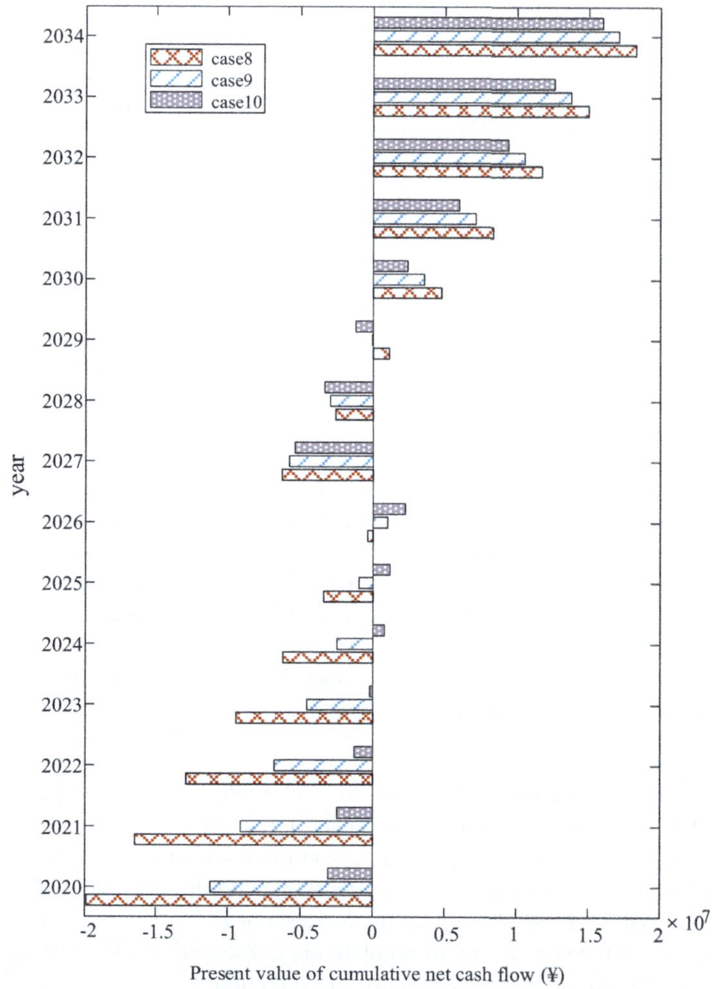

Fig. 3.10 Present value of C-NCF of case 8–case 10

3.4.2.3 Sensitivity Analysis for Uncertain Budgets

As stated in Sect. 3.1.3, the uncertain budgets denote the maximum number of periods when the PV generation and load demand are taken as the boundary of uncertain sets in the decision-making process. As a result, the uncertain budgets have important influence on the microgrid design scheme. Table 3.6 illustrates the influence of the uncertain budgets on the total cost and diesel fuel cost.

It can be seen that the total cost as well as the diesel fuel cost increases along with the increase of $\boldsymbol{\Gamma}^{\mathrm{load}}$ for a certain value of $\boldsymbol{\Gamma}^{\mathrm{PV}}$, as demonstrated in line 2–line 8, line 9–line 15, and line 16–line 22 of Table 3.6. Similarly, the total cost and the diesel fuel cost increase with the increase of $\boldsymbol{\Gamma}^{\mathrm{PV}}$ for a fixed $\boldsymbol{\Gamma}^{\mathrm{load}}$. The operational

Table 3.6 Total costs under different uncertain budgets

Item $\Gamma_{PV}/\Gamma_{load}$	Total cost ($\times 10^4$ ¥)	Diesel fuel cost ($\times 10^4$ ¥)
2/2	9332.77	5619.50
2/4	9480.01	5793.26
2/6	9610.78	5950.33
2/8	9733.43	6107.10
2/10	9830.91	6182.34
2/12	9929.28	6265.52
2/14	10075.24	6487.80
4/2	9442.39	6013.22
4/4	9602.51	6216.66
4/6	9746.07	6440.32
4/8	9868.77	6599.14
4/10	9973.08	6624.24
4/12	10074.88	6742.48
4/14	10102.27	6752.48
6/2	9516.24	6157.13
6/4	9663.02	6313.40
6/6	9784.45	6477.99
6/8	9893.09	6612.58
6/10	9990.04	6634.47
6/12	10083.35	6769.32
6/14	10167.80	6769.90

scenarios of the microgrid will be more “worse” with the increase of the uncertain budgets; thus, the DG generation has to raise to guarantee the load supply.

To sum up, the uncertain budgets will affect the realization of PV generation and load demand. The PV generation will be reduced with the increase of $\boldsymbol{\Gamma}^{\mathrm{PV}}$, and the load demand will increase along with the increase of $\boldsymbol{\Gamma}^{\mathrm{load}}$. Therefore, the uncertain budgets can be utilized as a way to regulate the conservatism of the design scheme according to the risk adverse reference of the stakeholders.

3.5 Conclusions

In this chapter, a dynamic microgrid planning and design model is established based on a typical scenario set generation method. The typical scenario set generation is formulated as an MIMLP problem that can realize a balance among multiple objectives and is able to extract both the statistical and time-series fluctuation information of the original data to guarantee the typicality and extremity. The dynamic microgrid planning and design model takes into account both the long-term uncertainty of fuel price and the short-term uncertainties of PV generation and load demand via hybrid stochastic programming and robust optimization techniques.

Numerous simulations have verified the effectiveness and superiority of the proposed model, and the major conclusions are summarized as follows:

1. The proposed model shows better performance in terms of total cost, PV utilization rate, and fund recovery speed over the traditional static design method. The investment decisions are more practical and economically attractive as the dynamic information such as load growth, unit cost of DERs, and device degradation are considered. Moreover, the advantages of the proposed model are more significant with the increase of load growth rate.
2. The proposed model can be used by the interested stakeholder to obtain the investment decision that is cost-effective and robust under multiple uncertainties. The conservatism of the design scheme can be regulated flexibly by adjusting the uncertain budgets to realize a balance between economy and robustness.
3. With the popularization of microgrid application, the demand of microgrid will result in an integration of electricity, heat, gas etc. Although only the electrical load is considered in this chapter, the proposed model is easily to be extended to an integrated energy framework to support the planning and design for integrated energy microgrids.

References

1. IEEE Recommended Practice for the Planning and Design of the Microgrid (2019)
2. B. Zhao, X.S. Zhang, P. Li, K. Wang, M.D. Xue, C.S. Wang, Optimal sizing, operating strategy and operational experience of a stand-alone microgrid on Dongfushan Island. Appl. Energy (Elsevier) **113**, 1656–1666 (2014)
3. P. Li, R.X. Li, Y. Cao, D.Y. Li, G. Xie, Multiobjective sizing optimization for island microgrids using a triangular aggregation model and the levy-harmony algorithm. IEEE Trans. Ind. Infor. (IEEE) **14**(8), 3495–3505 (2017)
4. A.A. Hamad, M.E. Nassar, E.F. El-Saadany, M.M.A. Salama, Optimal configuration of isolated hybrid AC/DC microgrids. IEEE Trans. Smart Grid (IEEE) ***10**(3)*, 2789–2798 (2018)
5. S. Mohamed, M.F. Shaaban, M. Ismail, E. Serpedin, K.A. Qaraqe, An efficient planning algorithm for hybrid remote microgrids. IEEE Trans. Sustain. Energy (IEEE) **10**(1), 257–267 (2018)
6. M. Rahmani-Andebili, Chapter 9: Cooperative distributed energy scheduling in microgrids, in *Electric Distribution Network Management and Control*, (Springer, Singapore, 2018), pp. 235–254
7. A. Khodaei, S. Bahramirad, M. Shahidehpour, Microgrid planning under uncertainty. IEEE Trans. Power Syst. (IEEE) **30**(5), 2417–2425 (2014)
8. L. Shi, Y. Luo, G.Y. Tu, Bidding strategy of microgrid with consideration of uncertainty for participating in power market. Int. J. Electr. Power Energy Syst. (Elsevier) **59**, 1–13 (2014)
9. L. Dusonchet, E. Telaretti, Comparative economic analysis of support policies for solar PV in the most representative EU countries. Renew. Sustain. Energy Rev. (Elsevier) **42**, 986–998 (2015)
10. M. Aien, A. Hajebrahimi, M. Fotuhi-Firuzabad, A comprehensive review on uncertainty modeling techniques in power system studies. Renew. Sustain. Energy Rev. (Elsevier) **57**, 1077–1089 (2016)

11. M. Rahmani-Andebili, Dynamic and adaptive reconfiguration of electrical distribution system including renewables applying stochastic model predictive control. IET Gener. Transm. Distrib. **11**(16), 3912–3921 (2017)
12. M. Rahmani-Andebili, Stochastic, adaptive, and dynamic control of energy storage systems integrated with renewable energy sources for power loss minimization. Renew. Energy **113**, 1462–1471 (2017)
13. M. Rahmani-Andebili, M. Fotuhi-Firuzabad, An adaptive approach for PEVs charging management and reconfiguration of electrical distribution system penetrated by renewables. IEEE Trans. Ind. Infor. **14**(*5*) (2018)
14. M. Rahmani-Andebili, Chapter 4: Robust operation of a reconfigurable electrical distribution system by optimal charging management of plug-in electric vehicles considering the technical, social, and geographical aspects, in *Planning and Operation of Plug-In Electric Vehicles: Technical, Geographical, and Social Aspects*, (Springer, Cham, 2019)
15. R.E. Moore, R.B. Kearfott, M.J. Cloud, Introduction to interval analysis, *Society for Industrial and Applied Mathematics,* (2009)
16. W. Prommee, W. Ongsakul, Optimal multiple distributed generation placement in microgrid system by improved reinitialized social structures particle swarm optimization. Eur. Trans. Electr. Power (Wiley) **21**(1), 489–504 (2011)
17. S.A. Arefifar, Y.A.I. Mohamed, T.H.M. EL-Fouly, Optimum microgrid design for enhancing reliability and supply-security. IEEE Trans. Smart Grid (IEEE) ***4***(*3*), *1567–1575* (2013)
18. S. Sukumar, M. Marsadek, A. Ramasamy, H. Mokhlis, Grey Wolf Optimizer Based Battery Energy Storage System Sizing for Economic Operation of Microgrid, *IEEE International Conference on Environment & Electrical Engineering & IEEE Industrial & Commercial Power Systems Europe (IEEE),* (2018)
19. P. Xie, Z. Cai, P. Liu, X.H. Li, Y.J. Zhang, D.J. Xu, Microgrid system energy storage capacity optimization considering multiple time scale uncertainty coupling. IEEE Trans. Smart Grid (IEEE) **10**(5), 5234–5245 (2018)
20. M. Kumar, B. Tyagi, An optimal multivariable constrained nonlinear (MVCNL) stochastic microgrid planning and operation problem with renewable penetration. IEEE Syst. J. (IEEE) **14** (3), 4143–4154 (2020)
21. R. Atia, N. Yamada, Sizing and analysis of renewable energy and battery Systems in Residential Microgrids. IEEE Trans. Smart Grid (IEEE) **7**(3), 1204–1213 (2016)
22. M. Rahmani-Andebili, Analyzing the Effects of Problem Parameters on the Operation Cost of the Networked Microgrids, 2020 *IEEE Kansas Power and Energy Conference (KPEC)*, (2020)
23. L. Guo, R.S. Hou, Y.X. Liu, C.S. Wang, H. Lu, A novel typical day selection method for the robust planning of stand-alone wind-photovoltaic-diesel-battery microgrid. Appl. Energy (Elsevier) **263**, 114606 (2020)
24. U.S. Energy Information Administration, [Online]. available: https://www.eia.gov/petroleum/gasdiesel/
25. E.L. Hannan, Linear programming with multiple fuzzy goals. Fuzzy Sets Syst. (Elsevier) **6**(3), 235–248 (1981)
26. Y.P. Cai, G.H. Huang, Z.F. Yang, Q. Tan, Identification of optimal strategies for energy management systems planning un-der multiple uncertainties. Appl Energy (Elsevier) **86**(4), 480–495 (2009)
27. E.S. Lee, R.J. Li, Fuzzy multiple objective programming and compromise programming with Pareto optimum. Fuzzy Sets Syst. (Elsevier) **53**(3), 275–288 (1993)
28. D. Bertsimas, E. Litvinov, X.A. Sun, J.Y. Zhao, T.X. Zheng, Adaptive robust optimization for the security constrained unit commitment problem. IEEE Trans. Power Syst. (IEEE) **28**(1), 52–63 (2013)

29. E. Vartiainen, G. Masson, C. Breyer, D. Moser, E.R. Medina, Impact of weighted average cost of capital, capital expenditure, and other parameters on future utility-scale PV levelised cost of electricity. Prog. Photovolt. Res. Appl. (Wiley) **28**(6), 439–453 (2020)
30. T. Adefarati, R.C. Bansal, Reliability, economic and environ-mental analysis of a microgrid system in the presence of renewable energy resources. Appl. Energy (Elsevier) **236**, 1089–1114 (2019)
31. X. Zhao, T.R. Brown, W.E. Tyner, Stochastic techno-economic evaluation of cellulosic biofuel pathways. Biores. Technol. (Elsevier) **198**, 755–763 (2015)
32. M.A. Syakur, B.K. Khotimah, E.M.S. Rochman, B.D. Satoto, Integration K-means clustering method and elbow method for identification of the best customer profile cluster. IOP Conf. Ser. Mater. Sci. Eng. (IOP) **336**, 12–17 (2018)

Chapter 4
Hierarchical and Distributed Dispatching of Microgrids Considering Uncertainty

Xiangyu Kong, Dehong Liu, Wenqi Lu, Chengshan Wang, Yu Shen, Wei Hu, and Mehdi Rahmani-Andebili

Abstract The interconnected microgrid system (IMS) is a promising solution for the problem of growing penetration of renewable-based microgrids into the power system. To optimally coordinate the operation of microgrids owned by different owners while considering uncertainties in market environment, a bi-level distributed optimized operation method for IMS with uncertainties is proposed in this chapter. A hierarchical and distributed operational communication architecture of IMS is first established. A bi-level distributed optimization model was built for IMS, where at the upper level, the IMS operates purchase-sale mode or demand response mode with the distribution network operator and optimizes the trading power with microgrids to maximize revenue. At the lower level, the chance constraint programming is used to describe and deal with the uncertainty of renewable energy and loads and optimize the output and energy storage of distributed energy with the goal of minimum cost. The analytical target cascading and augmented Lagrange method are combined to decouple and reconstruct the bi-level model for distributed solution and establish a fair price mechanism. The optimal solutions of the problem are obtained through parallel iteration, in which the price signal plays a coordinated role in the distributed iterative optimization process. Abundant case studies verify the advantages of the model and the performance of the proposed method.

Keywords Multi-microgrid system · Decentralized framework · Bi-level energy dispatch · Uncertainty · Power Internet of things

---

X. Kong (✉) · D. Liu · W. Lu · C. Wang
Key Laboratory of Smart Grid of Ministry of Education, Tianjin University, Tianjin, China
e-mail: eekongxy@tju.edu.cn

Y. Shen · W. Hu
State Grid Hubei Electric Power Research Institute, Hubei, China

M. Rahmani-Andebili
Department of Engineering Technology, State University of New York, Buffalo State, Buffalo, NY, USA
e-mail: rahmanm@buffalostate.edu

M. Rahmani-Andebili (ed.), *Design, Control, and Operation of Microgrids in Smart Grids*, Power Systems, https://doi.org/10.1007/978-3-030-64631-8_4

4.1 Introduction

4.1.1 Background

Microgrids (MGs) can flexibly and efficiently integrate and utilize distributed energy sources locally, which improves the reliability and energy efficiency of local power supply and reduces the impact of renewable energy fluctuations on the utility grid [1]. In recent years, the maturity of distributed renewable energy technologies and a new round of power system reform have promoted the emergence of a large number of MGs and independent energy service providers. As the penetration of these distributed devices increases, the control and operation of a large scale of MGs become exceptionally challenging. On one hand, the imbalance of peak and valley power and uncertainty of source and load have greatly affected the safe and reliable operation of a single MG. On the other hand, with the gradual opening of the electricity market and the decline of renewable energy subsidies, MG operators (MGOs) are actively seeking more profitable operation modes. The interconnected MG system (IMS) is a plurality of MGs aggregated through electrical and information connections in a particular area, which aims to promote the energy exchange of regional MGs, improve the level of renewable energy consumption, and ensure regional power safety and stability [2]. However, the source and load composition and the supply and demand characteristic of each MG are different, and they have different operation modes guided by their own interest targets [3]. With the goal of building an open and shared local energy system, IMS should not only meet the "immediate access, immediate operation" and dynamic exit of MGs but also meet the requirements for local user privacy and data security during operation. The former puts forward higher requirements for the communication and operation architecture of IMS, while the latter means that MG and IMS may only have limited information interaction, which cannot be met by traditional methods of centralized data storage and model solution. In addition, the uncertainty of renewable energy and loads and the problem of price mechanism also brought difficulties to system operation and energy trading.

4.1.2 Literature Review

At present, the researches on optimal operation of multiple MGs (MMGs) can be classified into two categories: integrated operator (IO) model and decentralized operator (DO) model. In IO model, the IMS as a whole unit shares a common goal of benefit, performing a centralized operation [4]. Nikmehr et al. [5] developed a centralized optimization control method for MG exchanging energy with the main grid to minimize the total cost of MMGs. Again, a centralized control model for optimal management and operation of MMGs is designed to obtain the optimal strategy of power flows in the network in [6] by Ouammi et al. who then proposed a cooperative control scheme for a smart network of residential buildings under uncertainties in the renewable sources and loads in [7] showing significant advantages and benefits of central cooperation of integrated MG with one master controller. Arefifar et al. in [8] presents an optimized strategy for performing energy management in IMS, the results of which shows that the centralized energy management is more beneficial compared to performing it separately for each MG. Generally, IO model can help IMS obtain the optimal operation strategy and fully enhance energy efficiency. The idea of the abovementioned centralized method is mainly to consider concentrating all the model data of IMS to form a complete network model. However, the model lacks scalability without considering the dynamic access and exit of MGs during actual operation. For large-scale IMS, this stepwise centralized scheduling method will cause excessive data and calculation burden on the calculation center, which has high requirements on the communication system and control system. In addition, in the future market environment, multiple operators are inevitable due to diverse business models [8], but the IO operation cannot guarantee the fair distribution of the benefits of each MG, which seriously weakens MG's motivation to access IMS.

Unlike IO, the DOs focus on solving the IMS operation problem in a decentralized way which is not only considered more suitable for multiple operators but also more consistent with the cloud-edge architecture of power Internet of things (PIoT). Advantages of a DO scheme include the ability to survive uncertain disturbances, as well as fully distributed data updating, which results in efficient privacy information sharing and, ultimately, faster decision-making and operation [9]. Among them, game theory provides a way to analyze the competition and cooperation between different operators. A game-theoretic noncooperative distributed coordination control scheme was developed by Liu et al. [10] to address multi-operator energy trading in multi-MGs. A bi-level model is proposed in [11] to find equilibria among MGs which are bidding/offering strategically to buy/sell power from/to the distributed network (DN) in a day-ahead market. The solutions of this EPEC are analyzed through a diagonalization method to identify meaningful Nash equilibria. The economic operation scheduling of distribution network operator in the presence of MMGs is formulated as a multi-follower game problem in [12] by defining MMGs and distributed network operator (DNO) as game players. The agent-based Nash bargaining strategy was employed to cooperatively obtain a fair and Pareto-optimal solution to the hierarchical power management problem of MMGs by Dehghanpour et al. [13]. An optimization strategy integrating a Stackelberg game to model the interactions between energy provider and consumers is proposed by Yang et al. [14] for distribution networks with MMGs. For most game theory-based approaches, the problem must be proven convex or transformed into a convex problem to guarantee the existence of equilibrium solutions. In addition, some game theory-based methods include a heterogeneous decomposition for the different stakeholders which is a sequential solution procedure [4]. MMG's consumption of computing resources rises rapidly with the increase of MG. Therefore, the hierarchical and distributed optimal scheduling is advanced in the control and management of IMS with multiple operators. A hierarchical decentralized energy management system was developed and further formulated as a bi-level optimization problem by Zhao et al. [15] to coordinate the operation of networked MGs in distribution systems where the power exchanges both within and between MGs were considered. Ref. [16] presents a multiagent-based hierarchical energy management strategy incorporation of DR to minimize operational costs of individual MGs and ensure the supply reliability of the IMS simultaneously. To address the day-ahead scheduling optimization of smart distribution networks with multiples MGs, a bi-level interactive mechanism was proposed in [17]. Later, novel design architecture and control for IMSs with three-level decentralized control were also presented by Wu et al. [18]. However, in the hierarchical structure, operators' decisions at the lower level depend heavily on upper-level decisions, which means that the computation of the upper and lower layers of each interaction is sequential, not parallel.

In addition, uncertainties in renewable energy and loads could make scheduling plans with deterministic information impractical or even collapse. Diverse uncertainties from renewable energy sources, power and heat loads, as well as power transaction prices are integrated into the system operation model and addressed by the multi-energy coordination by Li et al. [19]. The robust optimization (RO) is often used to handle the MMG operation with DER's uncertainties. Bertsimas et al. [20] proposed a two-stage adaptive robust unit commitment model for the security-constrained unit commitment problem. Inspired by two-stage adaptive RO, the author in [21] follow the similar line and present an advanced model considering the discrete characteristics of energy transaction combinations among MGs and employ the column-and-constraint generation algorithm to solve the problem efficiently. A stakeholder-parallelizing distributed adaptive robust optimization model is proposed by Qiu et al. [22] for the scheduling of hybrid AC/DC MMGs to realize the global scheduling of tie lines and finally determine a robust plan. However, the solution obtained by RO starting from the worst case is conservative. Ma et al. [23] present online energy management based on the online alternating direction method of multipliers (ADMM) algorithm which does not require any forecast for DERs and has less conservative schedule than the RO-based approach. But such a method is limited to short-period scheduling with online data, which is hard to solve the economic dispatch problem of IMS with off-line information. Different from RO, stochastic programming (SP) provides another way to handle uncertainty. The energy management of IMS in distribution systems is formulated as a stochastic bi-level optimization problem in [24]. Hybrid stochastic/robust optimization of an optimal bidding strategy for MG was presented in ref. [25], where the uncertain output of DGs was formulated by SP and the unbalanced power in the real-time market is limited by RO, which balanced the robustness and conservatism of the solution. Long-term uncertainty like the declining or increasing trend of storage investment cost and short-term uncertainty like renewable energy generation and load were considered by Wei et al. [26] to form a multi-period planning for multi-energy MGs. To coordinate the long-term and the short-term uncertainty, Wei et al. developed a chance-constrained-information-gap-decision-based model and then converted such model into a mixed-integer linear programming equivalence solved by commercial solver CPLEX released by IBM. In [27], the effects of problem parameters on the operation cost minimization of networked microgrids are investigated for the given planning period. To minimize the total cost of problem, the networked microgrids are optimally reconfigured, and the DGs and capacitor banks are optimally sized and installed in the best locations. In [28–31], a stochastic model predictive control (stochastic MPC) is applied to model the uncertainty and variability of renewable power and load demand. To sum up, most existing studies use robust optimization and stochastic programming to deal with the uncertainties of sources and loads. Whether RO or SP is the more appropriate way of handling uncertainty still depends on the application. The traditional RO method usually adopts a safe but poor economic strategy, which leads to a conservative solution result. In the context of PIoT, the explosion of intelligent terminal and the application of big data technology can further grasp the probabilistic characteristics of renewable energy and load uncertainty. In this context, stochastic programming method will play a greater advantage than robust optimization.

Discussions on uncertainty and distributed hierarchical multi-operator architectures were numerous and diverse. However, externally, the role of IMS in interacting with the utility grid and internally fair mechanism for the energy exchange among the MGs are not well addressed in these existing works. Most of the existing work focuses on the optimal dispatching of multi-MGs, but little discussed about the interaction between MMG system and the utility grid. With the development and construction of the PIoT, the problems below are still obscure in the current work, such as in what form is the MMG system interconnected with the utility grid, in what role is it interacting with the utility grid, and with what function to help improve the utility grid.

4.1.3 Contributions and Significance

This chapter aims to solve the problem of optimal operation of IMS in the market environment, promote energy trading of MGs in the region, and exploit benefits of stakeholders. In order to realize the optimal economic dispatch of multiple MGs, many existing literatures have proposed the hierarchical distributed optimization method. However, the following problems remain to be further addressed. First, although these methods can achieve accurate model of different stakeholders, such as MGs and aggregator, there are always coupling variables in the model, which makes them difficult to solve and hard to achieve parallel solutions for MG operators and aggregator during the optimization process, reducing the calculation efficiency and weakening the advantages of decentralization and autonomy of the MGs. Besides, most work said little about the communication architecture of IMS. Second, the uncertainty of DERs and load will directly affect the safe operation of MG and the energy transaction of IMS. Most works to solve the uncertainties of IMS focused on robust optimization, which usually adopts a strategy with high security but poor economy, resulting in the conservative solution. However, the enhancement of perception ability and the increase of data volume and dimension make us more accurately grasp the characteristics of the deviation of renewable energy forecast and load forecast. Third, the electricity price traded by operators is often set as a fixed electricity price by default, just as in [32], which obviously is not suitable for dynamic operational changes of MGs under uncertainty. Moreover, most of the authors focused on solving the difficulties in the internal optimal scheduling of IMS and talked briefly about the application that can be realized by the interaction between IMS and utility grid. The latter means more to IMS to realize social value and extend business.

To address these challenges, a bi-level distributed optimized operation method for IMS with uncertainties is proposed in this chapter. Firstly, a cloud-edge cooperative distributed operation communication architecture of interconnected MG system (IMS), combined with power Internet of Things, including the IntelliSense layer, edge computing layer, network transmission layer, and cloud platform application layer, is established. The cloud-edge coordination architecture enables IMS operator (IMO) and distribution network operator to interact in the cloud platform layer and formulate demand response (DR) plans. The edge computing of IMO and MGs achieves local optimization. Secondly, an IMS hierarchical and distributed optimal scheduling model is established in the market-oriented environment to maximize the interests of all stakeholders under the IMS cloud-edge architecture. The upper level considers the purchase sale or DR with DNO and coordinates the energy transaction between IMO and MGs, and the lower level considers the uncertainties of renewable energy and load and optimizes the internal resources dispatching of MG, like DERs and ESs. Finally, the model was decomposed based on the analytical target cascading (ATC) algorithm to adapt to the PIoT-based distributed communication computing architecture and improve the computing efficiency. The result shows that the energy trading between MGs and IMO was promoted to reduce their operational costs. Meanwhile, the utility grid benefits from peak load clipping. Based on the AL method, a fair price mechanism is designed as a coordinated signal for energy trading of IMO and MGs, which encourages MGs to participate in regional energy trading. The diagonal quadratic approximation (DQA) is employed for parallel solution of the upper- and lower-level procedure, which further shortens the time of solving the optimization operation of IMS. The impact of the different confidence level in dealing with uncertainty on stakeholder decision-making is systematically analyzed in case study. Based on the above contributions, the hierarchical and distributed optimal scheduling of IMS considering uncertainty studied in this chapter has important value and practical guiding significance, including the following points:

1. To promote the information physical fusion of regional distributed energy system. By integrating the architecture of IntelliSense layer, edge computing layer, network transmission layer, and cloud platform application layer, the cloud-edge collaborative architecture of IMS is established to realize real-time perception, information interconnection, and optimal control of regional distributed resources

and loads and promote the physical information fusion of regional distributed energy system.

2. To realize the efficient integration and utilization of various distributed resources in the region. MG aggregates regionally dispersed interactive resources such as DERs, controllable loads, and electric energy storage (ES). These resources are complex and diverse in types, capacity, quantities, and rights subjects and are confronted with multiple uncertainties of sources and loads.

The rest of this chapter is organized as follows: Sect. 4.2 establishes the operational and communication structure of the IMS. Section 4.3 built the mathematical model of bi-level optimal operation of IMS. In Sect. 4.4, the price mechanism is established, and parallel solution is achieved based on DQA method. Different case studies are evaluated in Sect. 4.5.

4.2 Operation Architecture Based on PIoT

Comprehensive and reliable communication technology is the foundation of IMS hierarchical and flexible scheduling. PIoT is an information physical network that is fully aware of the ubiquitous massive information in all links of the power system. Through the deep integration of big data, cloud computing, AI, 5G, and blockchain and other advanced information technologies with smart power grid, PIoT connects information islands and realizes the interconnection of devices and man-machine interaction in various business fields.

4.2.1 The Hierarchical Architecture of PIoT

According to the definition of the IEC/ISO, the Internet of things (IoT) is "an infrastructure connecting things, people, systems and information resources, which, combined with intelligent services, enables it to process and respond to information in the physical and virtual world" [33]. The early stages of IoT were a one-way flow of information, collecting data by installing sensors to connect physical objects but not by controlling the devices. With the development of intelligent terminal devices such as smart homes and electric vehicles, IoT cannot only collect operating data of devices but also give instructions to devices for real-time control. By integrating AI analysis and decision-making, IoT will greatly improve its intelligence level. In recent years, thanks to the rapid development of information technology, the concept of IoT has been widely developed and applied in various industries and fields.

PIoT is an industrial-grade IoT applied to power grids. Its layered architecture inherits the three layers of IoT's IntelliSense layer, edge computing layer, network

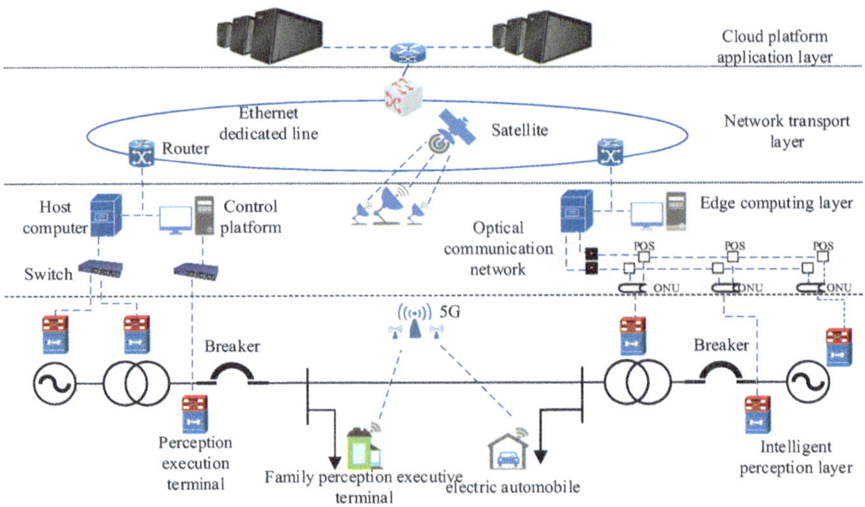

Fig. 4.1 The hierarchical architecture of PIoT

transmission layer, and cloud platform application layer and adds an edge computing layer [34], as shown in Fig. 4.1.

The IntelliSense layer is composed of numerous state perception and executive control main terminals, such as smart electricity meter, smart socket, home energy management terminal, and device sensor. It carries out "intelligent I/O" operations such as data acquisition, measurement, device control, and parameter adjustment [35], realizes the information acquisition and control of the underlying devices, and uploads the data to the edge computing layer. The edge computing terminal is a distributed intelligent agent that is close to the object data source and integrates the functions of network communication, data storage, computing and device management to provide intelligent services close to the clients. The network transmission layer undertakes the data transmission of multiple communication modes in a wide range between the edge computing layer and the cloud platform application layer. The network transmission layer consists of a local communication network and a telecommunication network. The local communication network realizes the connection between massive sensing nodes and edge computing nodes through flexible, efficient, and low-power near distance communication. The telecommunication communication network connects the edge computing layer and the cloud platform application layer with wide-area communication technologies such as high reliability and low delay satellite communication, electric power LTE network, and 5G mobile communication. The cloud platform application layer is the cloud computing center of PIoT, providing data storage, computing, and application software services, which is equipped with the unified management and processing capability of ultra-large data. It builds a data center platform for the traditional internal business of grid companies and carries out secondary application development of data information for external innovative business to realize advanced application.

4.2.2 Cloud-Edge Collaborative Operation Architecture of IMS

The emergence of PIoT mainly affects IMS from the following three aspects:

1. Enhanced the capability of communication and distributed collaborative computing between MGs, which provides a communication and computing foundation for MMG collaborative applications based on distributed interaction.
2. Enhanced the capability of state awareness of the IMS, including the ability to perceive basic data such as utility grid operation, renewable energy and load forecasting, market trading signals, and external environment. The probability distribution statistics of the forecast deviation have become more accurate due to the increase in the amount of data and the application of big data.
3. The IMS and other energy service entities become functional units in the cloud platform layer of grid company to realize advanced applications including DR and auxiliary services.

The increase of small independent energy systems such as MGs and virtual power plants breaks the traditional way that DNO manages all the DERs in the distribution network. The proposal of PIoT provides an interconnected architecture and interactive platform for these decentralized and independent energy systems. The hierarchical operation and communication architecture of IMS established in this chapter is shown in Fig. 4.2. Consider an MMG system consisting of a series of autonomous MGs and an aggregator in a regional distribution network. Each MG consists of DERs and users connected to other MGs through distribution lines and communication links. Usually, the MGO is the owner of the MG. Note that this chapter

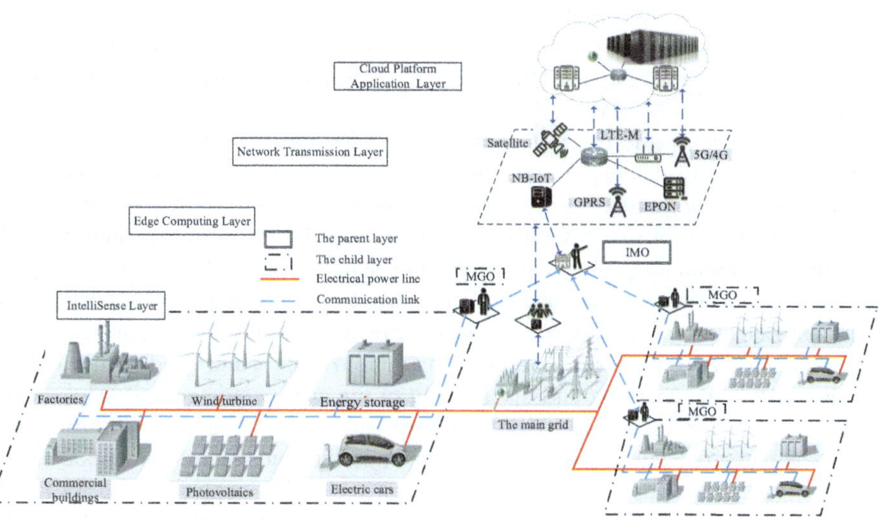

Fig. 4.2 The operation and communication structure of IMS

considers MG as a single interest entity and does not discuss the internal interest relationship of MG. The interconnected MG aggregator aggregates adjacent MGs in a regional distribution network and provides an energy trading platform to ensure fair energy trading among all participants.

On one hand, IMS internally coordinates MG energy exchange through optimal scheduling and pricing mechanism and creates profits. Distributed execution is carried out by each control platform of the edge computing layer, and multiple MG edge computing controllers are synergistically autonomous. On the other hand, IMS cannot only purchase and sell power from distribution network but also execute the DR plan of DNO. The purchase-sale mode and DR mode are two opposite operational paths of IMS where the difference lies in the role of tie line playing in the optimization process. The tie line can be regarded as a "virtue energy storage" absorbing the power surplus and deficit in purchase-sale mode while it is more like a "virtue load" having a power target requirement for IMS in DR mode. The uncertainty caused by prediction error is difficulty throughout the system. MGOs need to make a balance between the standby cost to deal with a certain degree of uncertainty and the penalty cost of generation deviation by IMS. A set of constraints including generator outputs, transmission flows, and consumption should be met in the energy trading.

4.2.3 Control Structure of MG

By relying on its own energy storage and controllable power supply and load, the MG coordinates and controls the distributed power supply, which solves the problem of large management dimension and complex control of the user-end DERs and helps to realize the maximization of energy efficiency and economic and environmental benefit. Operation control is the key to realize the above MG technology. At present, the research on operation control of MG is mainly divided into three layers [36].

The first layer is the inverter control for DERs and ES. The control goal is to respond to rapid load changes and maintain the stability of voltage and frequency. It belongs to the automatic control category of the equipment layer and usually adopts the drooping control mode with the response speed within milliseconds [37]. In addition to maintaining voltage and frequency stability, when the system is off-grid, the first layer (inverter control) mainly has the following two points in the interconnected microgrid system: (1) As the execution unit of the system scheduling plan, it controls the renewable energy such as photovoltaics and wind turbine output; (2) It controls the charging and discharging of the energy storage unit. The second layer is the research on mode switch control and voltage frequency control of the MG when off-grid. The control objectives include safe and controllable mode switch of the MG, as well as frequency and voltage regulation control of the MG after switching, which belong to the response control of the equipment layer [38]. The third layer is the research on optimal dispatching of MG energy management. The

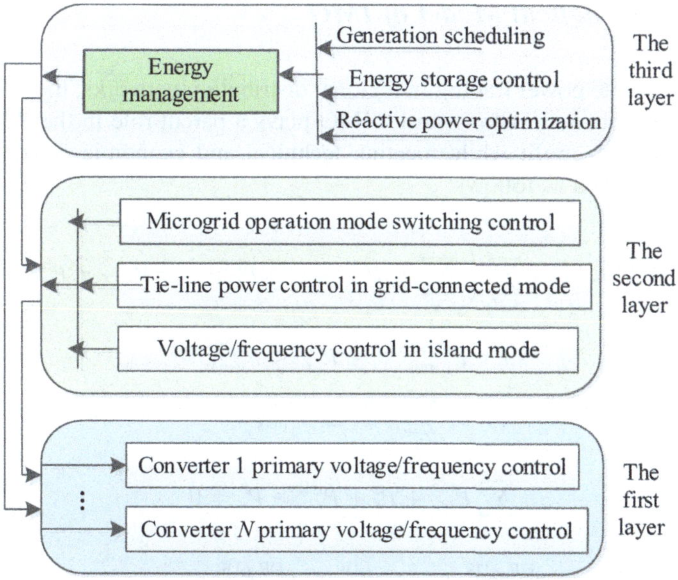

Fig. 4.3 Three control layers for MG

goal is to carry out dispatching control of DERs or ES in MG according to generation planning, reactive power optimization, and load forecasting results, to improve the economy of MG. This layer is realized by the upper computer and transmitted to the bottom inverter controller through the communication network. The three layers are coordinated with each other, and their structural relationship is shown in Fig. 4.3.

4.3 Hierarchical and Distributed Optimal Scheduling Method Considering the Uncertainties

In the process of economic dispatching of IMS, stakeholders like IMO and MGOs are driven by profit to pursue their own interests, which means that they have the objectives and constraints in the mathematical model of their own, which can be described as a modelling problem. Obviously, IMO is in the parent layer, and MG is in the child layer. However, due to the strong coupling between the parent layer and the child layer, the bi-level model is not only difficult to solve but also unable to adapt to the distributed computing architecture. In this chapter, a bi-level optimal operation model of IMS is built considering uncertainty, and the ATC method is employed to decouple the two layers.

4.3.1 Mathematical Model of IMO

IMO is to promote power trading in regional distribution networks, thereby improving power security and making profit. IMO plays a parent role in the IMS system, seeking maximum profit while meeting technical and economic constraints. The model is established as follows:

$$(\mathcal{P}) : F^{\mathrm{IMO}} = \max \sum_{t \in N_T} \left(\sum_{i \in N_{\mathrm{mg}}} \lambda_{i,t} P_{i,t}^{\mathrm{mg}} + \pi_t^{\mathrm{DR}} \left| P_t^{\mathrm{DR}} \right| + \pi_t^{\mathrm{s}} P_t^{\mathrm{s}} - \pi_t^{\mathrm{b}} P_t^{\mathrm{b}} \right), \tag{4.1}$$

$$\text{s.t.} \quad P_{i,\min}^{\mathrm{IMS} \leftrightarrow \mathrm{mg}} \le P_{i,t}^{\mathrm{mg}} \le P_{i,\max}^{\mathrm{IMS} \leftrightarrow \mathrm{mg}}, \tag{4.2}$$

$$\sum_{i \in N_{\mathrm{mg}}} P_{i,t}^{\mathrm{mg}} \le P_{t,\max}^{\mathrm{IMS}}, \tag{4.3}$$

$$\sum_{i \in N_{\mathrm{mg}}} P_{i,t}^{\mathrm{mg}} + P_t^{\mathrm{s}} + P_t^{\mathrm{DR}} - P_t^{\mathrm{b}} = 0 \ , \tag{4.4}$$

$$u_t^{\mathrm{DR}} P_{t,\max\_-}^{\mathrm{DR}} \le P_t^{\mathrm{DR}} \le u_t^{\mathrm{DR}} P_{t,\max\_+}^{\mathrm{DR}} \ , \tag{4.5}$$

$$0 \le P_t^{\mathrm{b}} \le u_t^{\mathrm{b}} P_{t,\max}^{\mathrm{IM} \leftarrow \mathrm{grid}} \ , \tag{4.6}$$

$$0 \le P_t^{\mathrm{s}} \le u_t^{\mathrm{s}} P_{i,\max}^{\mathrm{IM} \rightarrow \mathrm{grid}} \ , \tag{4.7}$$

$$u_t^{\mathrm{DR}} + u_t^{\mathrm{s}} + u_t^{\mathrm{b}} \le 1 \ . \tag{4.8}$$

Formula (4.1) is the objective function. T is the set of scheduling periods. There are 24 scheduling periods in a day in this chapter. $P_{i,t}^{\mathrm{mg}}$ is the trading power between IMS and MG i at time t. The positive value means that MG purchases power from IMO, and a negative value means the opposite. $\lambda_{i,t}^{\mathrm{mg}}$ is the price of energy exchange between IMO and MG at time t, which usually represents the marginal cost of energy transactions between IMS and MG. The pricing method will be introduced in Sect. 4.4. P_t^{DR} is the DR power of IMS to DNO at time t, and π_t^{DR} is the corresponding price. P_t^{s} and P_t^{b} are the purchasing and selling power of IMS to grid at time t. π_t^{s} and π_t^{b} are the corresponding prices. Note that according to different scenarios, the DR power can be positive or negative.

Formulas (4.2)–(4.8) are the constraints that the IMO economic dispatching model needs to meet. Formula (4.2) illustrates the power constraints of the tie line. Formula (4.3) restricts the total transaction power to prevent the multiple MG system from blocking. Formula (4.4) ensures power balance between buyers and sellers. Formula (4.5) constrains the power range of the DR. Formula (4.6) and Formula (4.7) restrict the trading power between IMO and the grid. Formula (4.8) denotes that purchasing and selling and DR cannot be executed at the same time.

4.3.2 Stochastic Optimization Model of MGs

Chance-constrained programming is a branch of stochastic programming, which is mainly used to solve the problem that the constraint condition contains random variables and the decision must be made before the realization of random variables is observed [39]. In consideration of the fact that some constraint conditions cannot be satisfied due to random variables in the actual situation, a confidence level is set, and the probability of the constraint conditions being established should be no less than this confidence level. A common form of chance-constrained programming is as follows:

$$\begin{aligned} &\min f(x) \\ &s.t.\ \Pr\{g_j(x,\xi) \le 0, j = 1, 2, \cdots, m\} \ge \alpha_j \end{aligned} \tag{4.9}$$

where x is the n-dimensional decision variables, $f(\mathrm{x})$ is the objective function, ξ is the random parameter vector, $g_j(x, \xi)$ is the random constraint function, $\Pr\{\cdot\}$ is the probability of event establishment, and α_j is the confidence level of the constraint conditions given in advance by the decision maker.

4.3.2.1 Uncertainty Analysis

The uncertainties in MG operation come from many aspects, the most important of which is the inaccuracy of the prediction of renewable energy and load. In this chapter, the output of photovoltaic (PV) and WT and the load are regarded as the sum of the predicted value and error value, in which the predicted values are the deterministic variables and the error values are the random variables, whose characteristics can be extracted from the historical data set. Formula (4.10) is the expression of renewable energy output and load in MG i at time t.

$$\begin{cases} P_{i,t}^{\mathrm{WT}*} = P_{i,t}^{\mathrm{WT}} + \xi_t^{\mathrm{WT}} \\ P_{i,t}^{\mathrm{PV}*} = P_{i,t}^{\mathrm{PV}} + \xi_t^{\mathrm{PV}} \\ P_{i,t}^{\mathrm{L}*} = P_{i,t}^{\mathrm{L}} + \xi_t^{\mathrm{L}} \end{cases} \tag{4.10}$$

where $P_{i,t}^{\mathrm{WT}*}$, $P_{i,t}^{\mathrm{WT}}$, and ξ_t^{WT} are the actual value, predicted value, and predicted error value of WT at time t, respectively. The same goes for the other two equations. Historical data show that the power output of WT and PV is approximately subject to Weibull distribution and Beta distribution, respectively. Load prediction is also affected by weather and social events. The prediction errors of them can be represented by normal distribution [40].

$$\begin{cases} \sigma_{i,t}^{\mathrm{WT}} = \rho_i^{\mathrm{WT}} P_{i,t}^{\mathrm{WT}} + \rho_{i,\mathrm{ins}}^{\mathrm{WT}} P_{i,\mathrm{ins}}^{\mathrm{WT}} \\ \sigma_{i,t}^{\mathrm{PV}} = \rho_i^{\mathrm{PV}} P_{i,t}^{\mathrm{PV}} + \rho_{i,\mathrm{ins}}^{\mathrm{PV}} P_{i,\mathrm{ins}}^{\mathrm{PV}} \\ \sigma_{i,t}^{\mathrm{L}} = \rho_i^{\mathrm{L}} P_{i,t}^{\mathrm{L}} \end{cases} \tag{4.11}$$

where $\sigma_{i,\ t}$ represents the standard deviation of the prediction error at time t. ρrepresents the correlation coefficient, which can be obtained from the historical data set. $P_{i,\mathrm{ins}}^{\mathrm{WT}}$ and $P_{i,\mathrm{ins}}^{\mathrm{PV}}$ are installed capacity of the WT and PV, respectively.

4.3.2.2 Uncertainty Analysis

The randomness of the prediction error is directly modeled as a chance constraint condition, which reduces the dimension of the uncertainty feasible region. The chance constraint is that the reserve capacity at a certain confidence level can cope with uncertainties of source and load. In this chapter, micro-turbine (MT) is reserved to resist the uncertainties from random RES and loads. In an emergency, renewable energy would be cut off. Therefore, the economic scheduling problem of MG can be described as follows: by controlling the output of controllable DERs like MT, the power charging or discharging of ES, and the power purchasing or selling of tie line, the total operating cost is minimized with the power balance constraint, equipment constraints, and standby constraints being met. It can be modeled as follows:

$$(C): F_i^{\mathrm{mg}} = \min \sum_{t \in T} \begin{pmatrix} \lambda_{i,t} P_{i,t}^{\mathrm{mg}} + c_i^{\mathrm{dis}} P_{i,t}^{\mathrm{dis}} + c_i^{\mathrm{ch}} P_{i,t}^{\mathrm{ch}} \\ + a_i^{\mathrm{MT}} \left(P_{i,t}^{\mathrm{MT}}\right)^2 + b_i^{\mathrm{MT}} P_{i,t}^{\mathrm{MT}} \end{pmatrix}, \tag{4.12}$$

$$\text{s.t.} \quad P_{i,\min}^{\mathrm{IMS} \leftrightarrow \mathrm{mg}} \le P_{i,t}^{\mathrm{mg}} \le P_{i,\max}^{\mathrm{IMS} \leftrightarrow \mathrm{mg}}, \tag{4.13}$$

$$P_{i,t}^{\mathrm{MT}} + P_{i,t}^{\mathrm{mg}} + P_{i,t}^{\mathrm{PV}} + P_{i,t}^{\mathrm{WT}} + P_{i,t}^{\mathrm{dis}} = P_{i,t}^{\mathrm{ch}} + P_{i,t}^{\mathrm{L}} \tag{4.14}$$

$$Soc_{i,t} = Soc_{i,t-1} + \left(P_{i,t}^{\mathrm{ch}} \eta_i^{\mathrm{ch}} - P_{i,t}^{\mathrm{dis}} / \eta_i^{\mathrm{dis}}\right) / E_i \ , \tag{4.15}$$

$$Soc_{i,\min} \le Soc_{i,t} \le Soc_{i,\max} \ , \tag{4.16}$$

$$Soc_{i,T_0} = Soc_{i,T_{24}} \ , \tag{4.17}$$

$$0 \le P_{i,t}^{\mathrm{ch}} \le u_i^{\mathrm{ch}} P_{i,\max}^{\mathrm{ch}} \ , \tag{4.18}$$

$$0 \le P_{i,t}^{\mathrm{dis}} \le u_i^{\mathrm{dis}} P_{i,\max}^{\mathrm{dis}} \ , \tag{4.19}$$

$$u_i^{\mathrm{ch}} + u_i^{\mathrm{dis}} \le 1 \ , \tag{4.20}$$

$$P_{i,\min}^{\mathrm{MT}} \le P_{i,t}^{\mathrm{MT}} \le P_{i,\max}^{\mathrm{MT}} \ , \tag{4.21}$$

$$P_{i,\mathrm{D}}^{\mathrm{MT}} \le P_{i,t}^{\mathrm{MT}} - P_{i,t\text{-}1}^{\mathrm{MT}} \le P_{i,\mathrm{U}}^{\mathrm{MT}} \ , \tag{4.22}$$

$$P_r\{P_{i,\max}^{\mathrm{MT}} - P_{i,t}^{\mathrm{MT}} \geq \xi_{i,t}^{\mathrm{WT}} + \xi_{i,t}^{\mathrm{PV}} - \xi_{i,t}^{\mathrm{L}}\} \geq \alpha_i, \tag{4.23}$$

$$P_r\{P_{i,t}^{\mathrm{MT}} - P_{i,\min}^{\mathrm{MT}} \geq \xi_t^{\mathrm{WT}} + \xi_t^{\mathrm{PV}} - \xi_t^{\mathrm{L}}\} \geq \beta_i, \tag{4.24}$$

Formula (4.12) is the objective function. $P_{i,t}^{\mathrm{ch}}$ and $P_{i,t}^{\mathrm{dis}}$ are the charging and discharging power of ES, respectively. c_i^{ch} and c_i^{dis} are the corresponding cost coefficients. $P_{i,t}^{\mathrm{MT}}$ means the output of the MT. a_i^{MT} and b_i^{MT} are the quadratic and linear cost coefficients of MT, respectively.

Formula (4.13) to Formula (4.24) are the constraints. Formula (4.13) describes the upper and lower limits of the trading power. Formula (4.15) describes the changing process state of charge (SoC). Formula (4.16) represents the SoC restriction. Formula (4.17) ensures the periodic scheduling of ES. Formulas (4.18) and (4.19) describe the charging and discharging power restrictions. Formula (4.20) denotes that charging and discharging cannot happen at same the time. Formula (4.21) restricts the upper and lower limits of MT. Formula (4.22) describes the ramp rate constraint of the MT. Formulas (4.23) and (4.24) are positive and negative rotation standby opportunity constraints, respectively, which indicates that the probability of positive and negative standby successfully coping with uncertainty is greater than α_i and β_i, respectively.

The random variables of prediction error of renewable energy resources and load make it impossible to directly solve the model. There are generally two methods for dealing with random chance constraints: the first is to perform random simulation experiments and to obtain approximate solutions to the problem through a large number of sampling iterations, which is time-consuming and slow to calculate [41]. The second is for some special random chance constraints, which is solved by converting it to the form of a deterministic equivalence class. The second method is adopted in this chapter.

$$P_{i,\max}^{\mathrm{MT}} - P_{i,t}^{\mathrm{MT}} \geq \Phi^{-1}(\alpha_i)\sqrt{\left(\sigma_{i,t}^{\mathrm{WT}}\right)^2 + \left(\sigma_{i,t}^{\mathrm{PV}}\right)^2 + \left(\sigma_{i,t}^{\mathrm{L}}\right)^2} \tag{4.25}$$

$$P_{i,t}^{\mathrm{MT}} - P_{i,\min}^{\mathrm{MT}} \geq \Phi^{-1}(\beta_i)\sqrt{\left(\sigma_{i,t}^{\mathrm{WT}}\right)^2 + \left(\sigma_{i,t}^{\mathrm{PV}}\right)^2 + \left(\sigma_{i,t}^{\mathrm{L}}\right)^2} \tag{4.26}$$

where $\Phi^{-1}(\alpha_i)$ represents the lower α_i sublocus of the standard normal distribution function.

4.3.2.3 Decoupling of Bi-level Models

It can be seen from the above model that the coupling variable of parent layer and child layers is trading powers $P_{i,t}^{\mathrm{mg}}$ which brings such disadvantages as too large scale to solve, the solution process cannot be distributed, and the system flexibility is poor. The ATC is a method to solve decentralized, hierarchical coordination problems which first decompose the entire system into several subsystems by creating new

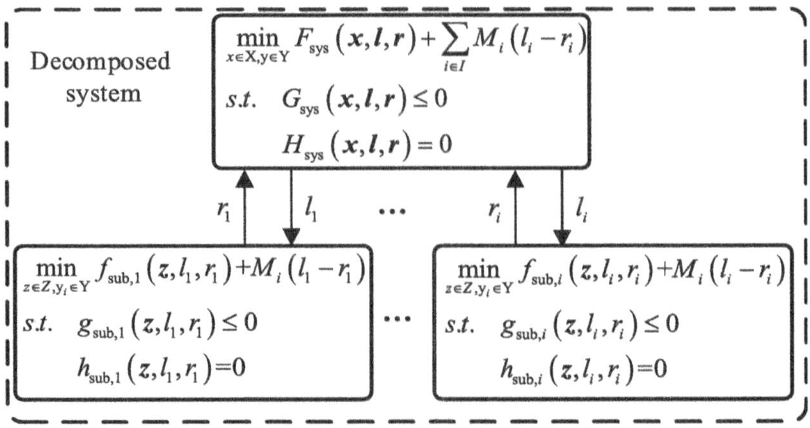

Fig. 4.4 The decomposition disgram of the bi-level model

shared variables and then design consistency constraints and additional penalty functions in each subsystem [42]. The basic idea is to solve the upper and lower problems independently and to optimize the overlap until the convergence condition is satisfied [43]. Take a typical hierarchical system A as an example:

$$(\mathcal{A}): \begin{cases} \min\limits_{x\in X,\, y\in Y} F_{\text{sys}}(\boldsymbol{x},\boldsymbol{y}) \\ s.t. \quad G_{\text{sys}}(\boldsymbol{x},\boldsymbol{y}) \le 0 \\ \qquad H_{\text{sys}}(\boldsymbol{x},\boldsymbol{y}) = 0 \\ \qquad \begin{cases} \min\limits_{z\in Z,\, y_i\in Y} f_{\text{sub},i}(z,y_i) \\ s.t. \quad g_{\text{sub},i}(z,y_i) \le 0 \\ \qquad h_{\text{sub},i}(z,y_i) = 0 \end{cases} \end{cases} \tag{4.27}$$

where $\boldsymbol{x}$ is the decision variables only related to the parent layer, $\boldsymbol{y}$ is the coupled variables of the parent layer and the child layers, and z is the decision variables only related to the child layer. Introduce target variables $\boldsymbol{l}$ to the parent layer and response variables $\boldsymbol{r}$ to the child layers to replace the coupling variables$\boldsymbol{y}$. The consistency constraint $l_i = r_i$ is added to ensure target/response consistency, reflected as a penalty function. The diverse penalty functions can increase the flexibility of the ATC method compared to other methods like ADMM [43]. Problem A is decoupled into problem B shown in Fig. 4.4.

The vector $\boldsymbol{r}$ is constant in the parent layer, and $\boldsymbol{l}$ is constant in the child layers. Each child layer is optimized independently, regardless of the connection between subsystems at the same level. The goal is to minimize the difference between the optimization result and the target provided by the previous layer. The inconsistencies of the optimization results of each subsystem are coordinated through the optimization of the upper system. Compared with other optimization methods, ATC has the advantages of parallel optimization, unrestricted series, and strict convergence proof. The proof of ATC's convergence and its application in multilevel optimization can be found in [44, 45], respectively.

Based on the ATC method, the target variable $P_{i,t}^{\text{imo}}$ and the response variable $P_{i,t}^{\text{mo}}$ were introduced to replace the coupling variable $P_{i,t}^{\text{mg}}$, and the original model $\mathcal{P}$ and $\mathcal{C}$ were transformed into $\mathcal{P}1$ and $\mathcal{C}1$ respectively.

$$(\mathcal{P}1):F^{\text{IMO}}=$$

$$\max\sum_{t\in T}\left\{\sum_{i\in N_{\text{mg}}}\lambda_{i,t}P_{i,t}^{\text{imo}}+\left(\pi_t^{\text{s}}P_t^{\text{s}}-\pi_t^{\text{b}}P_t^{\text{b}}\right)+\pi_t^{\text{DR}}\left|P_t^{\text{DR}}\right|-\sum_{i\in I}M_i\left(P_{i,t}^{\text{imo}},P_{i,t}^{\text{mo}}\right)\right\},\text{s.t. }(2)-(8). \tag{4.28}$$

$$(\mathcal{C}1):F_i^{\text{mg}}=\min\sum_{t\in T}\left\{\lambda_{i,t}P_{i,t}^{\text{mo}}+a_i^{\text{MT}}\left(P_{i,t}^{\text{MT}}\right)^2+b_i^{\text{MT}}P_{i,t}^{\text{MT}}+c_i^{\text{dis}}P_{i,t}^{\text{dis}}+c_i^{\text{ch}}P_{i,t}^{\text{ch}}+M_i\left(P_{i,t}^{\text{imo}},P_{i,t}^{\text{mo}}\right)\right\},$$
$$\text{s.t. }(13)-(22),(25),(26) \tag{4.29}$$

where $M_i(\cdot)$ is the penalty function related to $P_{i,t}^{\text{imo}}$ and $P_{i,t}^{\text{mo}}$ to coordinate the optimization process of the upper and lower layers. Note that the target variable and the response variable are not inverse of each other but gradually get closer with the iteration. The design $M_i(\cdot)$ and iteration rules are described in Sect. 4.4. In the decomposed model, it can be seen that local information, including prediction information and device parameter information, will not be uploaded to the parent layer. The only information MG and IMO interact with is the trading power information $P_{i,t}^{mg}$. This ensures the privacy of the MGs.

4.4 Price Mechanism and Parallel Solution

4.4.1 Price Mechanism of Energy Exchange Among MGs

The IMS distributed optimization model has many advantages, such as considering the demand-response interaction between IMS and DN and the uncertainties of MGs, and more importantly, it is a decoupled distributed structure which is very adaptable for the PIoT framework. Although there are many advantages of the bi-level optimization of MMG, we need to develop a price mechanism to ensure the trading fairness, which can thus motivate the MGs to share their surplus energy actively. In the proposed model in Sect. 4.3, the trading price of IMO and MG i at time t is a fixed value $\lambda_{i,\,t}$ without market meaning. Based on the Lagrange relaxation technology [46], the optimal Lagrange multiplier for each transmission line is the shadow price of the energy exchange in this transmission line, which indicates the

Table 4.1 The coordination process of the Lagrange multiplier as a price signal

MG i state	$P^{\text{mo}}_{i,t,k-1} - P^{\text{imo}}_{i,t,k-1} > 0$	$P^{\text{mo}}_{i,t,k-1} - P^{\text{imo}}_{i,t,k-1} < 0$
Purchase:$P^{\text{mg}}_{i,t} > 0$	High willingness to purchase, $\lambda_{i,t,k}$ rises	Low willingness to purchase, $\lambda_{i,t,k}$ falls
Sell:$P^{\text{mg}}_{i,t} < 0$	Low willingness to sell, $\lambda_{i,t,k}$ rises	High willingness to sell, $\lambda_{i,t,k}$ falls

marginal cost of the energy exchange. Inspired by this, the pricing mechanism was introduced into the IMS optimization model by designing the penalty function $M_i(\cdot)$ as augmented Lagrange form. The model introducing the price mechanism can be described as follows.

$$(\mathcal{P}2): F^{\text{IMO}} = \max \sum_{t \in T} \left\{ \sum_{i \in N_{\text{mg}}} \lambda_{i,t} P^{\text{imo}}_{i,t} + \left(\pi^{\text{s}}_t P^{\text{s}}_t - \pi^{\text{b}}_t P^{\text{b}}_t\right) + \pi^{\text{DR}}_t \left|P^{\text{DR}}_t\right| - \sum_{i \in N_{\text{mg}}} \begin{bmatrix} \lambda_{i,t}\left(P^{\text{imo}}_{i,t} - P^{\text{mo}}_{i,t}\right) \\ +\gamma_{i,t}\left(P^{\text{imo}}_{i,t} - P^{\text{mo}}_{i,t}\right)^2 \end{bmatrix} \right\},$$
$$\text{s.t. } (2)-(8). \tag{4.30}$$

$$(C2): F^{\text{mg}}_i = \min \sum_{t \in T} \left\{ \begin{array}{l} \lambda_{i,t} P^{\text{mo}}_{i,t} + c^{\text{dis}}_i P^{\text{dis}}_{i,t} + c^{\text{ch}}_i P^{\text{ch}}_{i,t} + a^{\text{MT}}_i \left(P^{\text{MT}}_{i,t}\right)^2 + b^{\text{MT}}_i P^{\text{MT}}_{i,t} \\ + \left[\lambda_{i,t}\left(P^{\text{mo}}_{i,t} - P^{\text{imo}}_{i,t}\right) + \gamma_{i,t}\left(P^{\text{mo}}_{i,t} - P^{\text{imo}}_{i,t}\right)^2\right] \end{array} \right\},$$
$$\text{s.t.} \quad (13)-(22), (25), (26). \tag{4.31}$$

where $\lambda_{i,t}$ is the Lagrange multiplier for the energy exchange between IMO and MG i at time t which at the same time is considered as the trading price indices to guide the energy trading and coordinate the power dispatch of IMO and MG i. The optimal values of Lagrange multiplier represent the marginal price of power trading between IMO and MG i. $\gamma_{i,t}$ is the coefficient of quadratic penalty term, increasing the convergence speed and the local convexity of the model to resist the influence of discrete variables in the model [47]. $\lambda_{i,t}$ and $\gamma_{i,t}$ are the coordination signals in the iterative solution process of the bi-level model, whose update rules can be written as follows according to the augmented Lagrange multiplier method [48].

$$\lambda_{i,t,k} = \lambda_{i,t,k-1} + 2\gamma_{i,t,k-1}\left(P^{\text{mo}}_{i,t,k-1} - P^{\text{imo}}_{i,t,k-1}\right) \tag{4.32}$$

$$\gamma_{i,t,k} = \beta\gamma_{i,t,k-1}, 1 < \beta \tag{4.33}$$

The initial value of the multipliers $\lambda_{i,t}$ and $\gamma_{i,j}$ are generally smaller constants. Note that in contrast to $\lambda_{i,t}$, the value of $\gamma_{i,j}$ has remarkable effect on convergence. If $\gamma_{i,j}$ is too small, then the optimization of MGs is unbounded. Moreover, large $\gamma_{i,j}$ may cause ill-conditioning problems [44]. Therefore, it is necessary to make trials for finding the proper parameter. The process $\lambda_{i,\ t}$ as a trading price feedback to coordinate energy trading can be simply described through Table 4.1.

Obviously, the optimal Lagrange multiplier of any MG may be different from others, which means that IMS implements differential pricing mechanism. The main reasons are as follows: (1) the cost coefficient of controllable DERs and ES is different in each MG; (2) the power supply and demand in each MG is different; (3) although the transmission loss is not included in this chapter, the line blocking constraint also has an impact on the price difference.

4.4.2 Parallel Solution

With ATC, the target/response variables are introduced to decompose the original model to realize the IMS bi-level distributed optimization operation with multiple entities. However, the optimization of the parent and child layers is not parallel. In the kth iteration, the result of the (k-1)-th iteration $P^{\text{mo}}_{i,t,k\text{-}1}$ is considered as fixed parameter to solve the model P2 to obtain $P^{\text{imo}}_{i,t,k}$, which then is used to solve the model C2 to obtain $P^{\text{mo}}_{i,t,k}$. It can be seen that the model is solved sequentially rather than in parallel, which reduces the computational efficiency of IMS optimization operation. In this chapter, the DQA method is applied to separate the penalty terms and achieve parallel computing for the parent layer and children layers [49]. The proof of feasibility and convergence of this method can be found in [50]. After approximation, the IMS bi-level distributed model can be rewritten as

$$
\begin{aligned}
&(\mathcal{P}3): F^{\text{IMO}}_{k+1} = \\
&\max \sum_{t\in T} \left\{ \sum_{i\in N_{\text{mg}}} \lambda_{i,t,k} P^{\text{imo}}_{i,t} + \left(\pi^{\text{s}}_t P^{\text{s}}_t - \pi^{\text{b}}_t P^{\text{b}}_t\right) + \pi^{\text{DR}}_t \left|P^{\text{DR}}_t\right| - \sum_{i\in N_{\text{mg}}} \begin{bmatrix} \lambda_{i,t,k}\left(P^{\text{imo}}_{i,t} - P^{\text{mo}}_{i,t,k}\right) \\ +\gamma_{i,t,k}\left(P^{\text{imo}}_{i,t} - P^{\text{mo}}_{i,t,k}\right)^2 \end{bmatrix} \right\}, \\
&\text{s.t. } (2)-(8).
\end{aligned}
\tag{4.34}
$$

$$
\begin{aligned}
(C3): F^{\text{mg}}_{i,k+1} = \min \sum_{t\in T} &\left\{ \begin{array}{l} -\lambda_{i,t,k} P^{\text{mo}}_{i,t} + c^{\text{dis}}_i P^{\text{dis}}_{i,t} + c^{\text{ch}}_i P^{\text{ch}}_{i,t} + a^{\text{MT}}_i \left(P^{\text{MT}}_{i,t}\right)^2 + b^{\text{MT}}_i P^{\text{MT}}_{i,t} \\ + \left[\lambda_{i,t,k}\left(P^{\text{mo}}_{i,t} - P^{\text{imo}}_{i,t,k}\right) + \gamma_{i,t,k}\left(P^{\text{mo}}_{i,t} - P^{\text{imo}}_{i,t,k}\right)^2\right] \end{array} \right\}, \\
\text{s.t.} \quad &(13)-(21), (24), (25).
\end{aligned}
\tag{4.35}
$$

The level-by-level sequential solution procedure is converted into the single-level parallel solution procedure. The algorithm process of IMS bi-level distributed parallel optimization based on DQA is as follows. The flowchart shown in Fig. 4.5 is presented to solve the proposed bi-level parallel optimization of IMS operation.

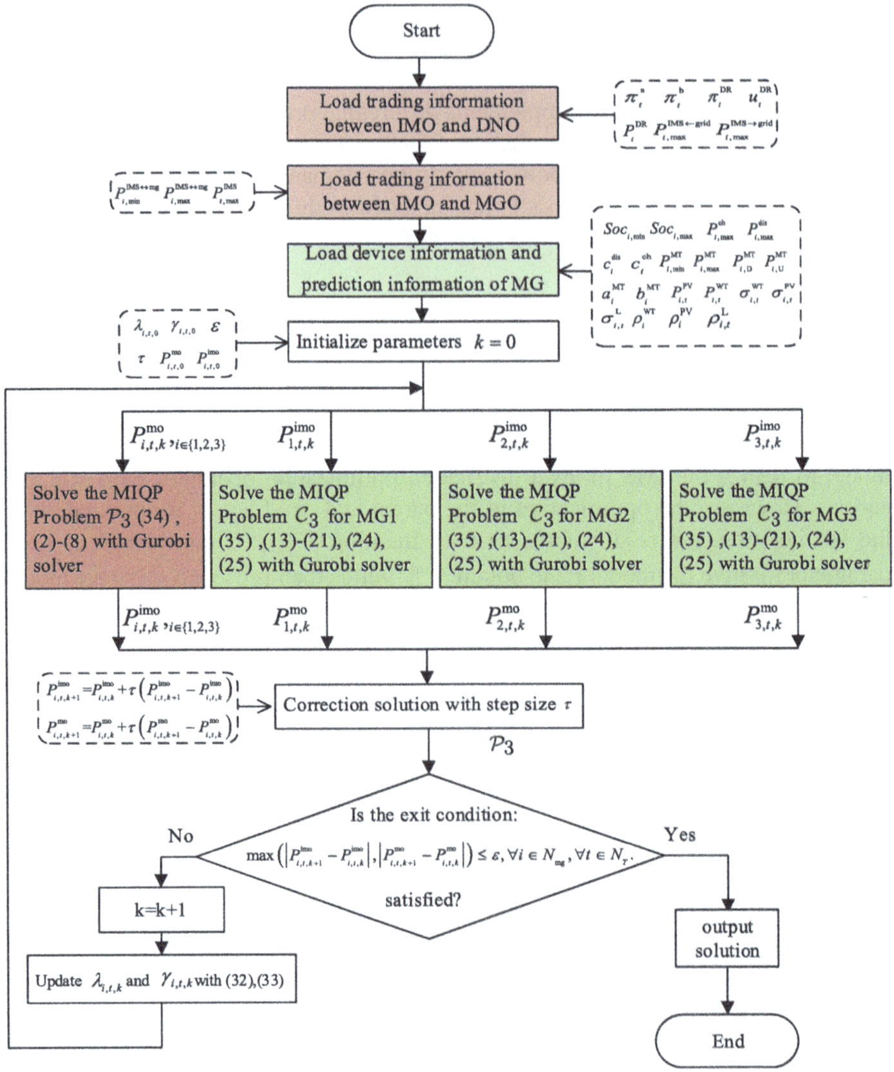

Fig. 4.5 Flowchart of ATC-based bi-level distributed parallel optimization

4.5 Case Study

In this section, detailed case studies are carried out to verify the effectiveness and performance of the bi-level distributed optimization method for the multi-stakeholder IMS under uncertainties. The code is programmed in MATLAB 2019 and solved by Gurobi solver, which operates on Intel® Core™ i5 1.99 GHz computer with 8 GB RAM.

4.5.1 Parameter Settings

The case studies are based on an IMS system consisting of three MGs and an IMO in a DN, as shown in Fig.4.1. The corresponding network can be found in [51]. The parameter settings between IMS and DN are as follows. $P_{t,\max}^{\text{IM}\leftarrow\text{grid}}$ and $P_{t,\max}^{\text{IM}\rightarrow\text{grid}}$ are set to 1500 kW and 1000 kW. Time-of-use (TOU) prices are used in purchasing and selling electricity from and to grid, shown in Table 4.2. The charging and discharging cost and efficiency of ES are 0.4 CNY/kWh and 0.6 CNY/kWh, and 0.95 and 0.97, respectively. All MGs are equipped with PV, WT, MT, ES, and loads, as well as calculation units for decision calculation. The specific configurations of MGs are shown in Table 4.3. The load and renewable energy forecasts in each MG are shown in Fig. 4.6. ρ_i^{WT}, $\rho_{i,\text{ins}}^{\text{WT}}$, ρ_i^{PV}, $\rho_{i,\text{ins}}^{\text{PV}}$, and $\rho_{i,t}^{\text{L}}$ are 0.15, 0.02, 0.1, 0.01, and 0.1, respectively. Convergence accuracy $\varepsilon$is set to 0.1. Initial values for $\lambda_{i,\ t,\ k}$ and $\gamma_{i,\ j,\ k\ -\ 1}$ are set to 0.27 and 0.5 CNY/kWh. Ignore the transmission loss in the IMS system. It is assumed that MGs can work in interconnected MG mode and island mode.

4.5.2 Comparison Between Grid-Connected Mode and Interconnected Mode

The grid-connected mode means the MG is directly connected to the DN without IMO and performs energy trading with DNO, not each other. Unlike the case of MG interconnection, direct trading between MG and grid is a single-layer optimization problem, which can be achieved by slightly changing model C. Based on the given configuration parameters, consider setting confidence level α and β to 0.8.

Figures 4.7 and 4.8 show the energy trading and electricity price result of MG1, MG2, and MG3 in both grid-connected and interconnected modes. The red step curve represents the price; the blue bar chart represents the power of MGs trading with IMO, where positive value means MG buying power from IMO and negative value means MG selling power to IMO. Figure 4.9 shows the sum of power exchange between the grid and MGs in two modes. Table 4.4 shows the cost of MGs and benefits of IMO in two modes. It can be seen in Figs. 4.7 and 4.8 that the transaction price curve of each MG is different from that of others. This is because

Table 4.2 TOU electricity price of the utility grid

Time	Price period	π^{b} (CNY/kWh)	π^{s} (CNY/kWh)
[1, 7]	Valley	0.405	0.279
[7, 8], [12, 18]	Average	0.702	0.484
[8, 10], [18, 19], [22, 24]	High	0.948	0.654
[10, 12], [19, 22]	Peak	1.096	0.756

Table 4.3 Configuration of MGs in IMS

MGs	P^{MT}_{min}(kW)	P^{MT}_{max}(kW)	P^{MT}_{U}(kW)	P^{MT}_{D}(kW)	a^{MT} ¥/kWh^2	b^{MT} ¥/kWh	E (kWh)	P^{ch}_{min}(kW)	P^{ch}_{max}(kW)	Soc_{min}	Soc_{max}	P^{tie}(kW)
MG1	65	500	300	−200	0.0005	0.7	400	120	200	0.2	0.9	750
MG2	60	500	250	−160	0.0005	0.7	300	90	150	0.2	0.9	750
MG3	50	400	240	−160	0.0005	0.7	500	150	250	0.2	0.9	500

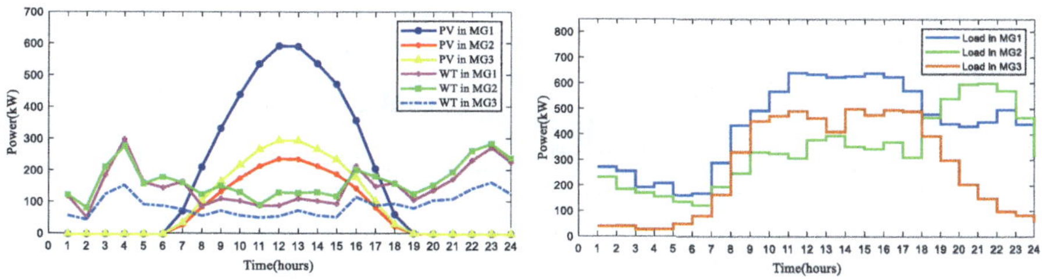

(a) (b)

Fig. 4.6 Hourly forecast for (**a**) renewable resources and (**b**) loads in MG1, MG2, and MG3

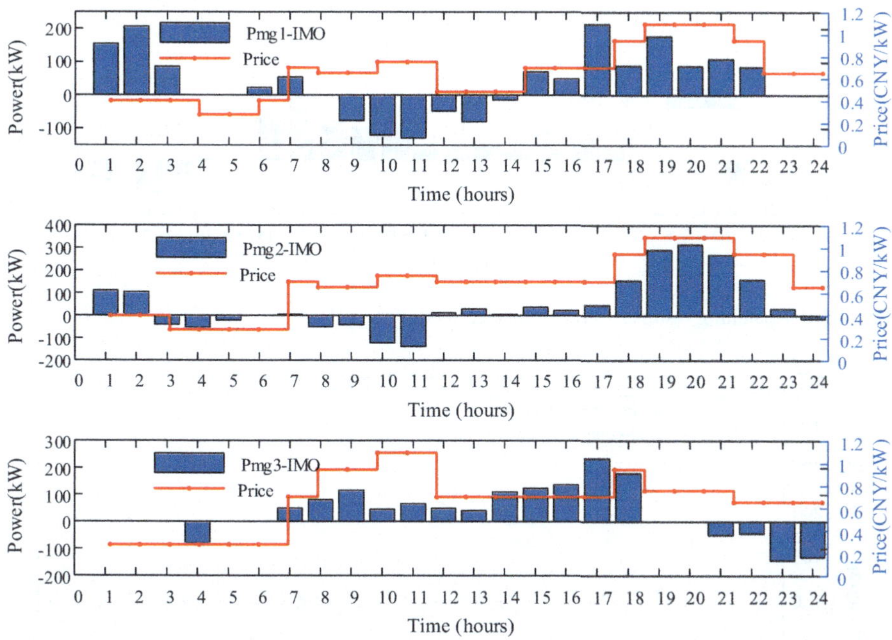

Fig. 4.7 Dispatching results of grid-connected mode

the supply and demand characteristics within the MG are different. For each MG, the transaction price represents the marginal cost of supply-demand balance with uncertainty. In grid-connected mode, the marginal cost of the local supply-demand balance of the MG is fixed to the TOU electricity price that is directly traded with the grid. In interconnected mode, IMO integrates the supply and demand of MMGs to form a larger local market. Each MG can get energy not only from the grid but also from other MGs under a fair price mechanism. Energy trading balance for all MGs is built by IMO, ensuring that local energy is fully absorbed and traded. On the other hand, it means that there is competition relationship among prices of MGs. As shown in Fig. 4.9, the volume of energy trading between MMGs and grids in the

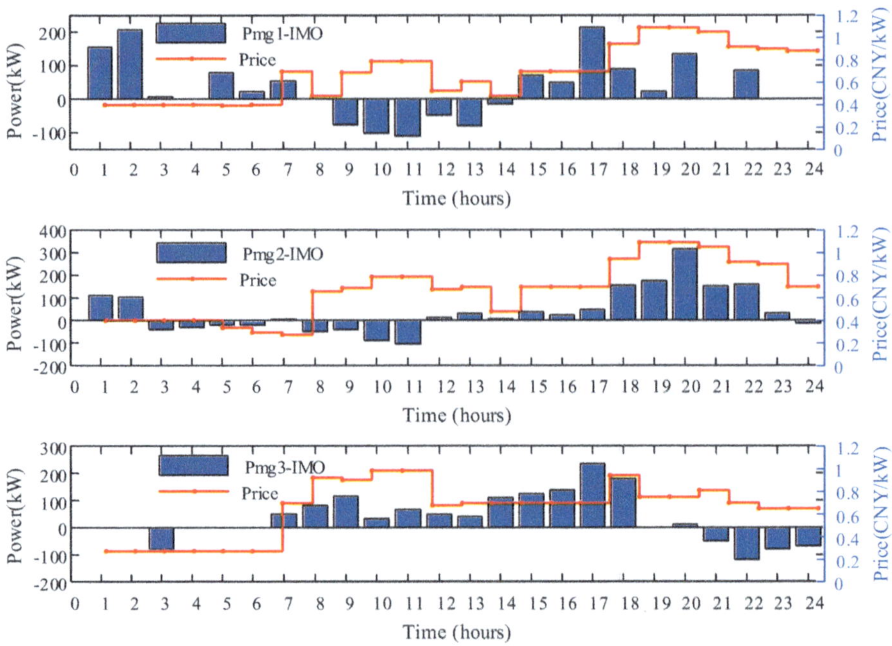

Fig. 4.8 Dispatching results of grid-connected mode

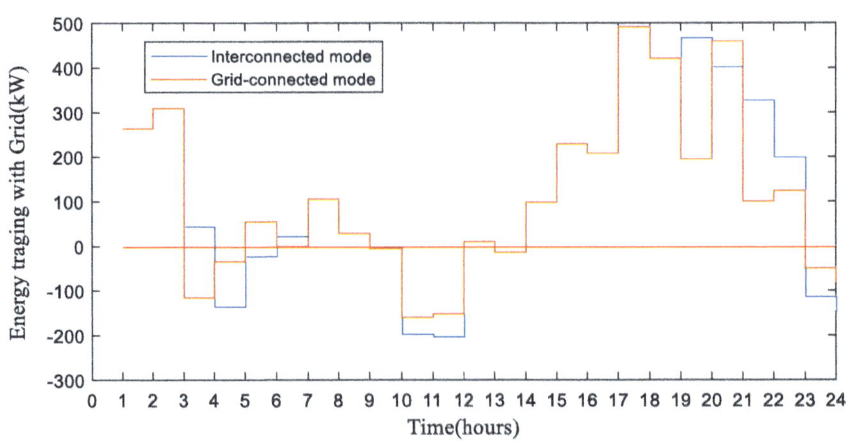

Fig. 4.9 Comparison of energy trading between the grid and MMGs in two modes

Table 4.4 Cost of MGs and benefit of IMO in two modes

Mode	MG1 (CNY)	MG2 (CNY)	MG3 (CNY)	IMO (CNY)
Grid-connected	1924.1	2088.2	1657.9	–
Interconnected	1905.8	2067.8	1582.5	76.37
Cost reduction %	−0.95%	−0.98%	−4.55%	*

interconnected mode is less than that in the grid-connected mode during hours 10–12 and hours 19–22. The case in the interconnected mode exchanges more power to the grid during hours 3–5. The trading prices are dramatically decreasing during peak price, promoting energy trading between MGs and IMOs to reduce their operational costs. IMO improves local energy utilization and reduces overall power supply costs. For the grid, meanwhile, it is obviously beneficial to peak load clipping of grid that IMS purchases less power during peak price period and sells more power during valley price period. In Table 4.4, the costs of MGs in the interconnected mode with IMOs can be seen reduced to different degrees, compared to that in the grid-connected mode. The costs of MG1, MG2, and MG3 were reduced by 0.95%, 0.98%, and 4.55%, respectively. The stakeholder of each MG could get benefits from the energy trading with IMOs. This is because IMO can reduce the price of energy purchased by MGs (lower than π^{buy} from grid) and increase the price of energy sold by MG (higher than π^{sell} from grid) through internal competitive transactions among MGs. Table 4.4 indicates that the proposed hierarchical optimization of interconnected MGs can effectively reduce the cost of MGs with a fair price mechanism.

4.5.3 *Operation of IMS in Different Uncertain Scenarios*

According to Formula (4.11), the parameter ρ in the model reflects the accuracy of the prediction and also reflects the severity of the fluctuation. The smaller the value, the higher the prediction accuracy, and the smaller the fluctuation. The impacts of forecasting accuracy and confidence level α on IMS scheduling results are analyzed. The IMS will operate in four uncertain scenarios with different prediction accuracy and confidence levels as shown in Table 4.5.

Figures 4.10 and 4.11 show the IMS scheduling result in scenario I to scenario IV. Table 4.6 shows the cost of MG and the benefit of IMO under the four operation scenarios of IMS with uncertainties. As the confidence level in the decision model increases, the reserve capacity required by the MGs increases. It can be seen from Figs. 4.10 and 4.11 that after the confidence level is increased, the MT's power generation has decreased overall. And the volumes and frequency of energy trading with IMO during the high-price time at night have increased; however, the amount of electricity sold to IMO is reduced. This change is more pronounced during periods when electricity consumption or renewable energy generation is high because the forecast deviation is positively related to the magnitude of the predicted value. When the confidence level is 0.9 in scenario II, the MG will not choose to start the MT to

Table 4.5 Uncertainty scenarios of MGs

	Prediction error coefficients $[\rho^{WT},\rho^{PV},\rho^{L}]$	
Confidence	[0.15, 0.1, 0.1]	[0.3, 0.2, 0.2]
$\alpha = 0.7$	Scenario I	Scenario III
$\alpha = 0.9$	Scenario II	Scenario IV

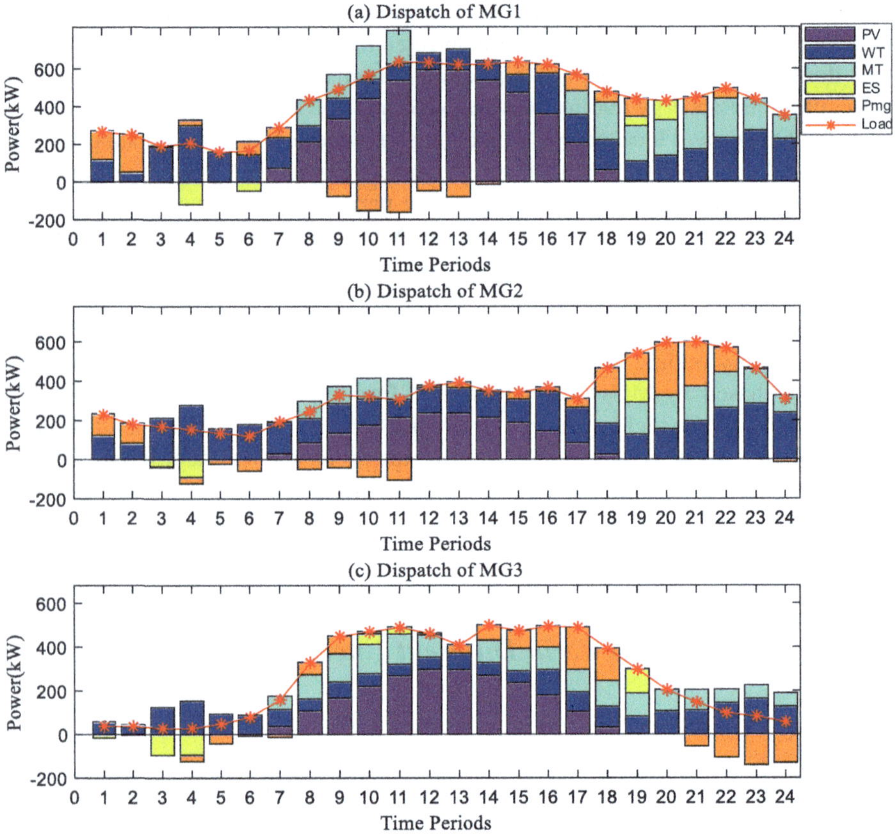

Fig. 4.10 Energy dispatch results of MGs in scenario I

sell electricity for profit at the high-price hours like 0.7 in scenario I, just as the behavior change of MG1 during hours 9–11. According to Table 4.6, as the confidence level increases, the costs of MGs increase and the revenue of IMO decreases. This is because the trading strategy of MGs becomes more conservative, and the complementary trading of MGs in IMS is reduced. Therefore, when making decisions, comprehensive consideration should be given to the reliability of the power supply and the economics of operation, and the confidence level should be selected appropriately. In practical application, confidence levels α and β are independently determined by stakeholders of each MG according to their conservatism of decision-making, which promotes the autonomy of the MGs. As prediction accuracy decreases and uncertainty increases, MG's stakeholders must keep more reserves and purchase more energy to make up for these deviations. As shown in Figs. 4.10 and 4.12, as the prediction accuracy decreases, the power generation of the controllable unit decreases, which is the same trend as the result of increasing the confidence level. However, increasing the confidence level from 0.7 to 0.9 has more impact on MG scheduling than double the prediction error coefficient.

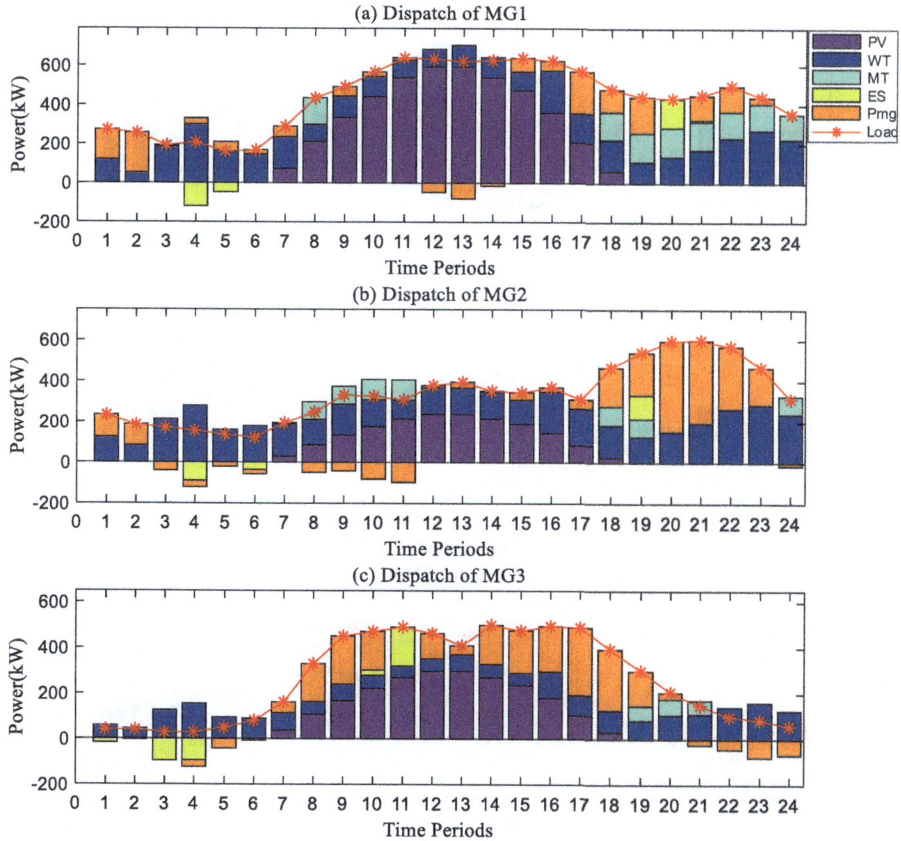

Fig. 4.11 Energy dispatch results of MGs in scenario II

Table 4.6 Cost of MGs and benefit of IMO in four scenarios

Scenarios	MG1 (CNY)	MG2 (CNY)	MG3 (CNY)	IMO (CNY)
Scenario I	1823.5	2012.9	1587.8	108.65
Scenario II	1960.9	2110.5	1668.5	56.29
Scenario III	1933.5	2033.4	1655.2	68.71
Scenario IV	2266.2	2347.6	1852.9	9.76

Table 4.6 tells that the cost of MGs for uncertainty in scenario III is lower than that in scenario II, while the benefits of IMO are opposite. In Fig. 4.13, the low accuracy of prediction and requirement of high confidence level make the MTs in MGs only used in small volumes at high power prices or low predicted values. Table 4.6 shows that the economics of MGs and IMO are very poor in scenario IV which illustrates the need for accurate forecasts of renewable energy and loads under PIoT development trends. In practical applications, large amounts of data generated by the IntelliSense layer and techniques such as machine learning can help reduce

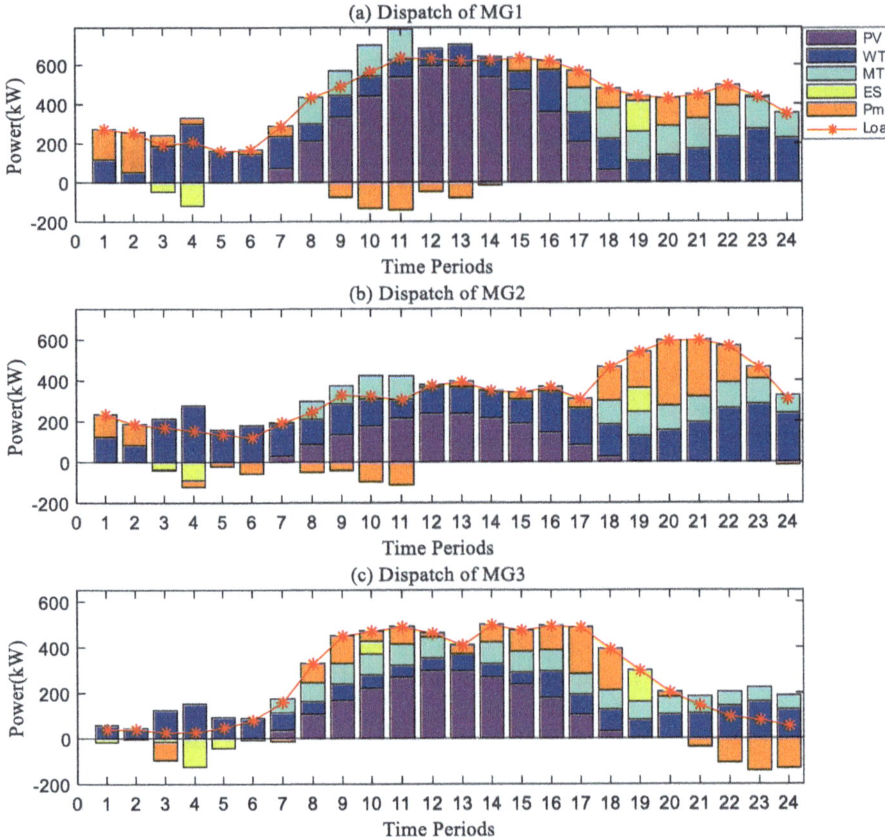

Fig. 4.12 Energy dispatch results of MGs in scenario III

prediction errors. The above analysis shows that the proposed method can help multi-stakeholders of IMS to adjust the dispatching scheme to deal with the impact of various uncertain scenarios while carrying out energy trading with a fair price mechanism.

4.5.4 IMS Implements Contract DR

According to the model $\mathcal{P}3$, IMS has two operating modes: purchase-sale (P-S) mode and DR mode. The operational capabilities of the IMS to participate in the DN contract DR are studied in this part. The DR implementation process in [52] mainly includes the following: IMS declares the DR amount, DNO notifies the DR period one day ahead, and IMS executes the contract DR volume. With reference to the "Tianjin Summer Demand Response Implementation Rules," the incentive

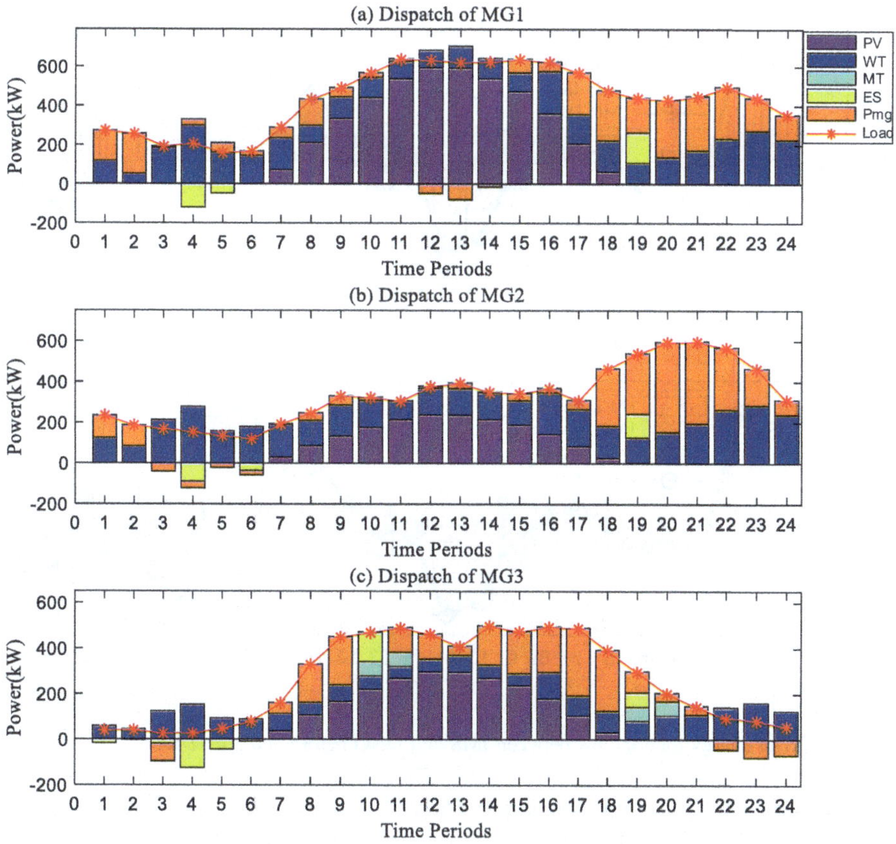

Fig. 4.13 Energy dispatch results of MGs in scenario IV

compensation for implementing the contract DR π^{DR} is 1.2 CNY/kWh. Assume that the DR period is notified as peak-price hours 19–21 by DNO. The declaration response volume of IMS is 400 kW. Take the scheduling result of scenario I as the baseline for examining the DR effect, as shown in Fig. 4.14a. The settings of the uncertainty environment are the same as scenario I.

Figure 4.14 shows the energy transactions between IMS and DNO before and after DR. Figure 4.15 shows the energy scheduling of the IMS in DR mode. IMO pre-sets the power of the tie line at the agreed time, which reduces the base load power by 400 kW. When the exchange power between IMO and DNO is fixed, the transaction power and energy scheduling of IMO and MGs will be affected. Higher DR incentive electricity price is used to change MG's power generation behavior as shown in Fig. 4.15. It is obvious that relative to the baseline, the IMO can fully execute the agreed DR volume at time 19–21. It is noteworthy that due to the MT climbing power limitation, the power purchase of MG3 in hour 18 decreased, which means that MG3 increased the MT power to meet the power requirement of MT in

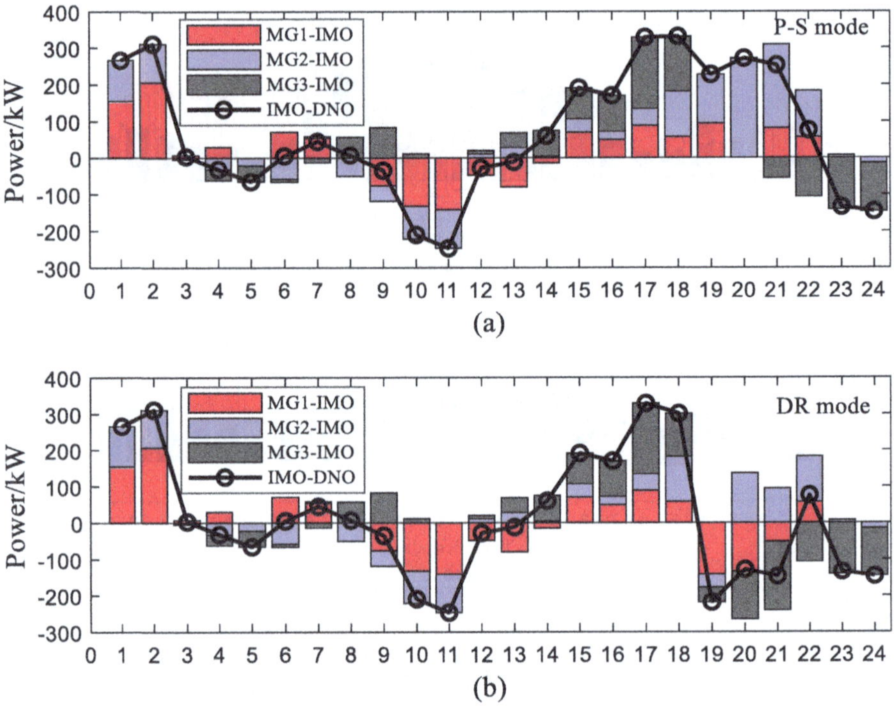

Fig. 4.14 The energy transactions between IMS and DNO before and after DR

hour 19 under the climbing constraint. The result shows that the proposed method can help IMS achieve the agreed DR to DNO and obtain more profits, which improves the application capabilities of IMS interaction with the public grid and realizes the wider social value of IMS.

4.5.5 Convergence and Efficiency of the Proposed Method

This section shows the convergence and efficiency of the proposed bi-level distributed optimization algorithm for IMS. The hierarchical distributed algorithm runs under the parameters described in Sects. 4.5.1 and 4.5.2. We observed the convergence process of the shared variables between MG1, MG2, and MG3 and IMO, as shown in Fig. 4.16. Figure 4.16 (a)–(c) and (d)–(f) show the convergence process of the target variable and the response variable of the parent and children models of IMS at period 10 and period 17, respectively. It can be seen that the independently optimized model $\mathcal{P}3$ and model $\mathcal{C}3$ converge to the same value within ten iterations under the coordination of the augmented Lagrange penalty term. It reflects the process of MG and IMO interactive optimization until the convergence conditions are met. Note that the time period 1–24 is optimized simultaneously in the model.

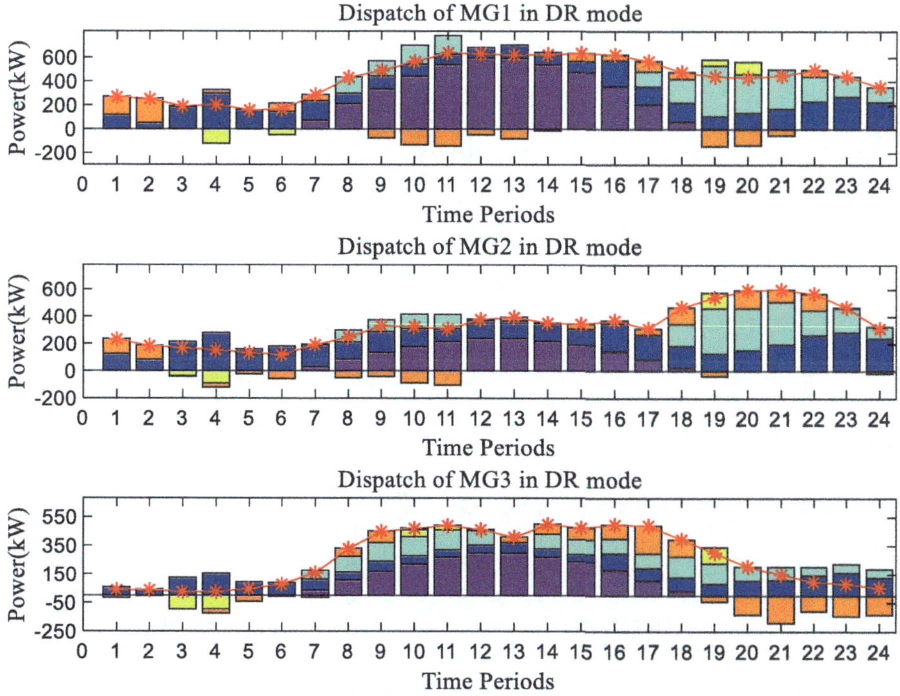

Fig. 4.15 Energy dispatch results of MGs in DR mode

To observe the computational efficiency of the proposed method, the energy scheduling problems of MMGs are solved using three kinds of algorithms, which are the interconnected mode with proposed method (ATC + AL + DQA), the interconnected mode with genetic algorithm (GA) solver, and the grid-connected mode with GA solver. It should be noted that the GA solver here comes from the built-in "GA" toolbox in MATLAB 2019. And instead of solving the whole bi-level model directly, GA solver solves the upper and lower levels of the bi-level model and exchanged shared variables. In the grid-connected mode, GA directly solves the energy trading problems between grid-connected MG1, MG2, MG3, and DNO.

As shown in Fig. 4.17 and Table 4.7, the interconnected mode converges to a lower cost than the grid-connected mode, the reason being explained in Sect. 4.5.2. Compared with GA, the proposed method (ATC + AL + DQA) has more iterations and shorter solution time for the interconnected mode. This means that each iteration of GA consumes more computational resources (related to parameters such as population size). In large-scale IMS applications, the proposed method is more applicable than GA algorithm. The grid-connected mode solved by GA can converge within three iterations because the model is simplified to a single-layer model, which is a small-scale mixed-integer quadratic programming problem. As a result, the method proposed in this chapter can achieve hierarchical distributed energy transaction optimization in IMS with an acceptable computing efficiency, and it

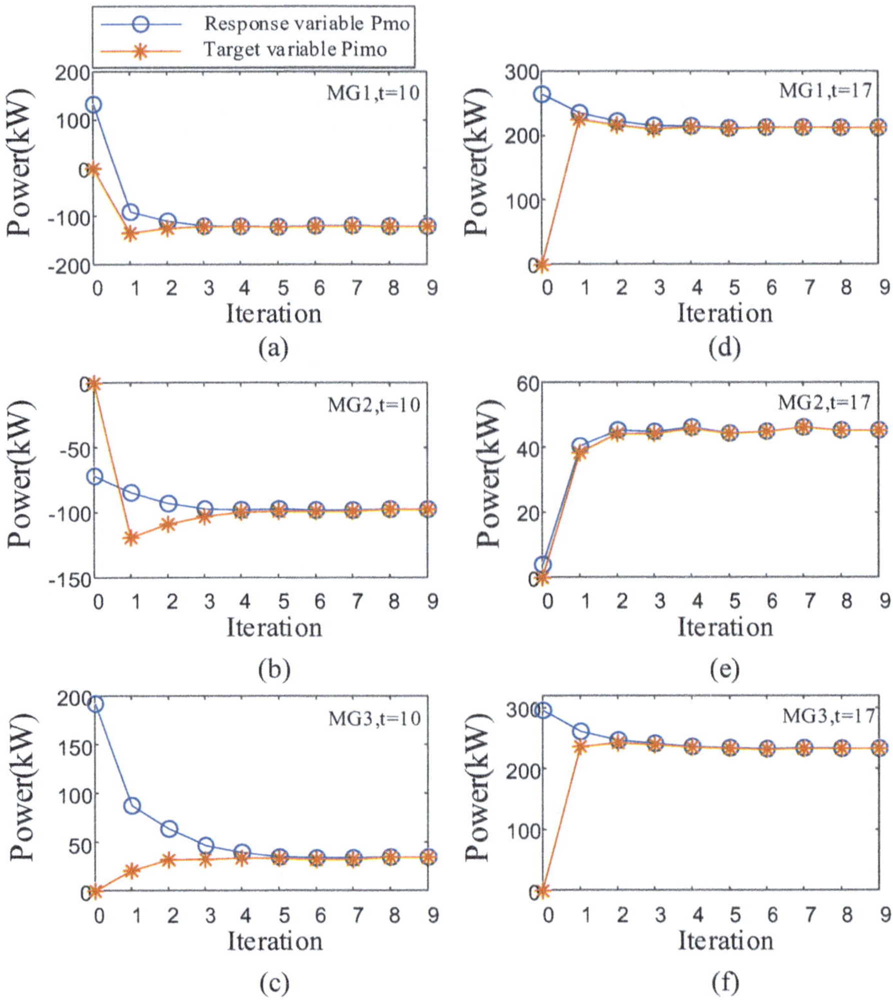

Fig. 4.16 The convergence process of IMS shared variables in: (**a**)–(**c**) period 10, (**d**)–(**f**) period 18

has a strong application value in combination with the cloud-edge collaborative PIoT architecture described in Sect. 4.2.

4.6 Conclusions

This chapter proposed a bi-level distributed optimized operation method for interconnected MG system with uncertainties to optimally coordinate the operation of MGs owned by a multi-stakeholder in market environment. The proposed

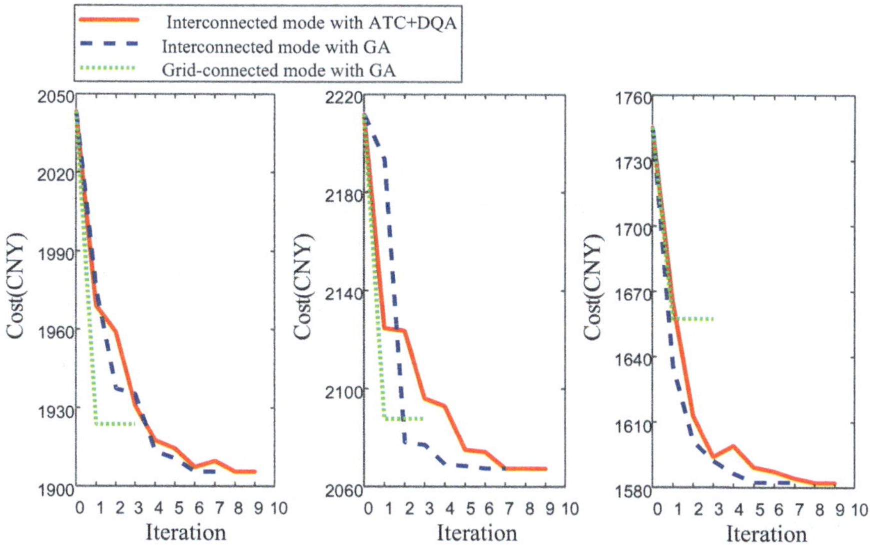

(a) Cost convergence of MG1 (b) Cost convergence of MG2 (c) Cost convergence of MG3

Fig. 4.17 The comparison of convergence process of cost: (**a**) MG1, (**b**) MG2, and (**c**) MG3

Table 4.7 Computational times of the different methods

Algorithm	Number of iterations	Solution time
Interconnected mode with ATC + AL + DQA	9	42.751 s
Interconnected mode with GA solver	7	248.335 s
Grid-connected mode with GA solver	3	31.561 s

interconnected MG system energy operational and communication architecture under power Internet of things supports the coordinated operation of MGs and integrated MG system operator while ensuring autonomy and privacy of MGs. The cloud-edge coordination architecture enables integrated MG system operator and DNO to interact in the cloud platform layer and formulate demand response plans. The edge computing of integrated MG system operator and MGs achieves local optimization.

The case study shows that DNO achieves cloud-edge coordination by influencing the behaviors of integrated MG system through time-of-use price and agreed demand response price. Based on this, DNO can also develop more applications in the cloud platform that can adapt to the diversified business mode of the power market, which has good scalability in the future applications. In order to evaluate the effectiveness of the proposed method, simulation studies were carried out in different test cases. Result shows that through implementing the proposed method, 0.95–4.55% reduction in cost of MG was achieved in the interconnected MG mode compared to grid-connected mode. The impact of renewable energy and load forecasting accuracy and confidence levels on the system cost has also been tested. The results showed that the method can help multi-stakeholders of integrated MG system to adjust the dispatching scheme to deal with the impact of various uncertain scenarios while carrying out energy trading driven by price. The simulation study of the convergence and efficiency of the proposed method shows that the cost of MG can converge to a lower value, and the system can achieve hierarchical distributed energy transactions within an acceptable time, which, in combination with the cloud-edge collaborative power Internet of things architecture, has a strong application value.

Finally, in this chapter, the loss between the integrated MG system and the MG is not taken into account which means the trading price among integrated MG system operator and MGs does not include power loss information. In addition, under the proposed framework, integrated MG system will have more applications to adapt to the diversified operating model of the power market. Future work will focus on these issues.

References

1. Z. Xu, P. Yang, C. Zheng, Y. Zhang, J. Peng, Z. Zeng, Analysis on the organization and development of multi-microgrids. Renew. Sust. Energ. Rev. **81**, 2204–2216 (2018)
2. E. Bullich-Massagué, F. Díaz-González, M. Aragüés-Peñalba, F. Girbau-Llistuella, P. Olivella-Rosell, A. Sumper, Microgrid clustering architectures. Appl. Energy **212**, 340–361 (2018)
3. L. Ren, Y. Qin, Y. Li, P. Zhang, B. Wang, P.B. Luh, S. Han, T. Orekan, T. Gong, Enabling resilient distributed power sharing in networked microgrids through software defined networking. Appl. Energy **210**, 1251–1265 (2017)
4. L. Wang, B. Zhang, Q. Li, W. Song, G. Li, Robust distributed optimization for energy dispatch of multi-stakeholder multiple microgrids under uncertainty. Appl. Energy **255**, 113845 (2019)
5. N. Nikmehr, S. Najafi Ravadanegh, Optimal power dispatch of multi-microgrids at future smart distribution grids. IEEE Trans. Smart Grid **6**, 1648–1657 (2015)
6. A. Ouammi, H. Dagdougui, R. Sacile, Optimal control of power flows and energy local storages in a network of microgrids modeled as a system of systems. IEEE Trans. Control Syst. Technol. **23**, 128–138 (2015)
7. A. Ouammi, Optimal power scheduling for a cooperative network of smart residential buildings. IEEE Trans. Sustain. Energy **7**, 1317–1326 (2016)
8. S.A. Arefifar, M. Ordonez, Y. Mohamed, Energy management in multi-microgrid systems – Development and assessment. IEEE Trans. Power Syst. **2**, 910–922 (2017)
9. L. Ma, N. Liu, L. Wang, J. Zhang, J. Lei, Z. Zeng, C. Wang, M. Cheng, Multi-party energy management for smart building cluster with PV systems using automatic demand response. Energ. Buildings **121**, 11–21 (2016)
10. W. Liu, W. Gu, J. Wang, W. Yu, X. Xi, Game theoretic non-cooperative distributed coordination control for multi-microgrids. IEEE Trans. Smart Grid, 1–1 (2017)
11. A. Naebi, S. SeyedShenava, J. Contreras, C. Ruiz, A. Akbarimajd, EPEC approach for finding optimal day-ahead bidding strategy equilibria of multi-microgrids in active distribution networks. Int. J. Electr. Power Energy Syst. **117**, 105702 (2020)
12. M. Jalali, K. Zare, H. Seyedi, Strategic decision-making of distribution network operator with multi-microgrids considering demand response program. Energy **141**, 1059–1071 (2017)
13. K. Dehghanpour, H. Nehrir, An agent-based hierarchical bargaining framework for power management of multiple cooperative microgrids. IEEE Trans. Smart Grid **10**, 514–522 (2019)
14. Y. Yang, W. Pei, Q. Huo, J. Sun, F. Xu, Coordinated planning method of multiple micro-grids and distribution network with flexible interconnection. Appl. Energy **228**, 2361–2374 (2018)

15. B. Zhao, X. Wang, D. Lin, M.M. Calvin, J.C. Morgan, R. Qin, C. Wang, Energy management of multiple microgrids based on a system of systems architecture. IEEE Trans. Power Syst. **33**, 6410–6421 (2018)
16. V. Bui, A. Hussain, H. Kim, A multiagent-based hierarchical energy management strategy for multi-microgrids considering adjustable power and demand response. IEEE Trans. Smart Grid **9**, 1323–1333 (2018)
17. Y. Liu, L. Guo, C. Wang, A robust operation-based scheduling optimization for smart distribution networks with multi-microgrids. Appl. Energy **228**, 130–140 (2018)
18. P. Wu, W. Huang, N. Tai, S. Liang, A novel design of architecture and control for multiple microgrids with hybrid AC/DC connection. Appl. Energy **210**, 1002–1016 (2018)
19. Z. Li, Y. Xu, Temporally-coordinated optimal operation of a multi-energy microgrid under diverse uncertainties. Appl. Energy **240**, 719–729 (2019)
20. D. Bertsimas, E. Litvinov, X.A. Sun, J. Zhao, T. Zheng, Adaptive robust optimization for the security constrained unit commitment problem. IEEE Trans. Power Syst. **28**, 52–63 (2013)
21. B. Zhang, Q. Li, L. Wang, W. Feng, Robust optimization for energy transactions in multi-microgrids under uncertainty. Appl. Energy **217**, 346–360 (2018)
22. H. Qiu, W. Gu, Y. Xu, Z. Wu, S. Zhou, G. Pan, Robustly multi-microgrid scheduling: Stakeholder-parallelizing distributed optimization. IEEE Trans. Sustain. Energy **11**, 988–1001 (2020)
23. W. Ma, J. Wang, V. Gupta, C. Chen, Distributed energy management for networked microgrids using online ADMM with regret. IEEE Trans. Smart Grid **9**, 847–856 (2018)
24. Z. Wang, B. Chen, J. Wang, M.M. Begovic, C. Chen, Coordinated energy management of networked microgrids in distribution systems. IEEE Trans. Smart Grid **6**, 45–53. 2015-01-01 2015
25. G. Liu, Y. Xu, K. Tomsovic, Bidding strategy for microgrid in day-ahead market based on hybrid stochastic/robust optimization. IEEE Trans. Smart Grid **7**, 227–237 (2016)
26. J. Wei, Y. Zhang, J. Wang, X. Cao, M.A. Khan, Multi-period planning of multi-energy microgrid with multi-type uncertainties using chance constrained information gap decision method. Appl. Energy **260**, 114188 (2020)
27. M. Rahmani-Andebili, Analyzing the effects of problem parameters on the operation cost of the networked microgrids, in *2020 IEEE Kansas Power and Energy Conference (KPEC)*, 2020
28. Chapter 4: Robust operation of a reconfigurable electrical distribution system by optimal charging management of plug-in electric vehicles considering the technical, social, and geographical aspects, in *Planning and Operation of Plug-In Electric Vehicles: Technical, Geographical, and Social Aspects* (Springer, Cham, 2019)
29. An adaptive approach for PEVs charging management and reconfiguration of electrical distribution system penetrated by renewables. IEEE Trans. Ind. Inf. 14(5) (2018)
30. Dynamic and adaptive reconfiguration of electrical distribution system including renewables applying stochastic model predictive control. IET Gener. Transm. Distrib. 11(16), 3912–3921 (2017)
31. Stochastic, adaptive, and dynamic control of energy storage systems integrated with renewable energy sources for power loss minimization. Renew. Energy. 113, 1462–1471, (2017)
32. X. Zhou, Q. Ai, M. Yousif, Two kinds of decentralized robust economic dispatch framework combined distribution network and multi-microgrids. Appl. Energy **253**, 113588 (2019)
33. I.I. JTC, *Internet of Things (IoT)* (Cham, International Organization for Standardization, 2014)
34. Z. Yan, L. Zeng, C. Xi, X. Li, J. Liu, S. Li, Architecture of power internet of things. Electricity **22**, 10–15 (2011)
35. X. Jiang, Y. Liu, F. U. Xiaofei, X. U. Peng, S. Wang, G. Sheng, Construction ideas and development trends of transmission and distribution equipment of the ubiquitous Power Internet of Things. High Voltage Eng. (2019)
36. T. Dragicevic, J.M. Guerrero, J.C. Vasquez, D. Skrlec, Supervisory control of an adaptive-droop regulated DC microgrid with battery management capability. IEEE Trans. Power Electron. **29**, 695–706 (2013)

37. S. Adhikari, F. Li, Coordinated V-f and P-Q control of solar photovoltaic generators with MPPT and battery storage in microgrids. IEEE Trans. Smart Grid **5**, 1270–1281 (2014)
38. J. Kirtley, H.H. Zeineldin, A.H.K. Alaboudy, Simple control strategy for inverter-based distributed generator to enhance microgrid stability in the presence of induction motor loads. Iet Gener. Transm. Distrib. **7**, 1155–1162 (2013)
39. J. Wang, An introduction to stochastic programming. Chinese J Oper Res **1**, 22–27 (1984)
40. J. Zhu, P. Duan, M. Liu, Electric real-time balance dispatch via bi-level coordination of source-grid-load of power system with risk. Proc. CSEE **35**, 3239–3247 (2015)
41. B. Liu, Y. Kai, *Uncertain Multilevel Programming* (Berlin/Heidelberg, Springer, 2015)
42. Y. Li, Z. Lu, J. Michalek, Diagonal quadratic approximation for parallelization of analytical target cascading. J. Mech. Des. **130**, 680–682 (2008)
43. A.R. Malekpour, A. Pahwa, Stochastic networked microgrid energy management with correlated wind generators. IEEE Trans. Power Syst. **32**, 3681–3693 (2017)
44. N. Michelena, H. Park, P. Papalambros, Convergence properties of analytical target cascading. AIAA J. (2003)
45. N. Kang, M. Kokkolaras, P.Y. Papalambros, S. Yoo, W. Na, J. Park, D. Featherman, Optimal design of commercial vehicle systems using analytical target cascading. Struct. Multidiscip. Optim. **50**, 1103–1114 (2014)
46. X. Guan, P.B. Luh, H. Yan, J.A. Amalfi, An optimization-based method for unit commitment. Int. J. Electr. Power Energy Syst. **14**, 9–17 (1992)
47. J.B. Lassiter, M.M. Wiecek, K.R. Andrighetti, Lagrangian coordination and analytical target cascading: solving ATC-decomposed problems with Lagrangian duality. Optim. Eng. **6**, 361–381. 2005-01-01 2005
48. S. Tosserams, L.F.P. Etman, P.Y. Papalambros, J.E. Rooda, An augmented Lagrangian relaxation for analytical target cascading using the alternating direction method of multipliers. Struct. Multidisc. Optim. **31**, 176–189 (2006)
49. C. Zhang, X. Lin, Analytical target cascading based on the quadratic exterior penalty method for complex system design. Adv. Mater. Res. **544**, 164–169 (2012)
50. Ruszczyński, Andrzej, On convergence of an augmented Lagrangian decomposition method for sparse convex optimization. Math. Oper. Res **20**, 634–656 (1995)
51. F.S. Gazijahani, S.N. Ravadanegh, J. Salehi, Stochastic multi-objective model for optimal energy exchange optimization of networked microgrids with presence of renewable generation under risk-based strategies. ISA Trans. **73**, 100–111 (2018)
52. H. Haddadian, R. Noroozian, Multi-microgrid-based operation of active distribution networks considering demand response programs. IEEE Trans. Sustain. Energy **10**(4), 1804–1812 (2019)

Chapter 5
Operation Strategy of Park Microgrid with Multi-stakeholder Based on Artificial Immune System

Xiangyu Kong, Dehong Liu, Fangyuan Sun, Chengshan Wang, Xianxu Huo, and Shupeng Li

Abstract The penetration of distributed energy resources (DER) is growing worldwide, and microgrid (MG) is an approprate way to realize intergration of these DERs. The new reform of power system promotes the market-oriented operation of microgrids. This chapter takes the park microgrid with multi-stakeholder as the object, and to promote the interaction between the main grid and DERs in MG, a two-level optimization model of microgrid bidding transaction based on multi-agent system is established. In the lower-level optimization, considering the deviation penalty of power generation and the previous round bidding results, the optimal bidding strategy model is established to maximize the benefit of bidding unit agent. In the upper-level model, bidding strategies of DERs as constraints, a multiple objective mixed-integer linear programming model was built to optimize the overall objectives of clearing price and imbalanced deviation, searching for the optimal clearing price and the generation plan of DERs. Due to the complexity of the two-layer optimization model, a novel artificial immune system (AIS) was established and integrated into the multi-agent system to help DERs participate in the optimal bidding operation of MG. The antigen is transformed by the environmental information, the price of the main grid, other DERs' bidding strategies, and the predicted deviation coefficient while considering the uncertainties of DER facilities. The proposed optimized operation mode is compared with the traditional operation mode in the case study, verifying that the proposed method can realize the optimal operation of the MG and the coordinated interaction with the main grid, increasing the benefit of stakeholders. The AIS algorithm is also compared with traditional algorithms, proving the superiority in optimizing.

Keywords Microgrid · Distributed energy · AIS · Bidding · Power market

X. Kong (✉) · D. Liu · F. Sun · C. Wang
Key Laboratory of Smart Grid of Ministry of Education, Tianjin University, Tianjin, China
e-mail: eekongxy@tju.edu.cn

X. Huo · S. Li
Tianjin Electric Power Company Electric Power Science Research Institute, Tianjin, China

M. Rahmani-Andebili (ed.), *Design, Control, and Operation of Microgrids in Smart Grids*, Power Systems, https://doi.org/10.1007/978-3-030-64631-8_5

5.1 Introduction

5.1.1 Background

With the continuous growth of population and the rapid development of economy and social productivity, the total demand for energy keeps increasing worldwide, but the traditional fossil energy is increasingly exhausted, and the world energy crisis is constantly staged. In addition, the burning of coal, oil, and other fossil fuels has caused serious environmental pollution, which has been troubling countries all over the world. The dual problem of energy shortage and environmental pollution urges all countries [1], on one hand, to actively seek renewable and clean energy solutions to replace traditional energy sources and improve energy utilization efficiency. On the other hand, it has become an international consensus to promote market-oriented energy reform, enhance market vitality, improve energy and resource allocation efficiency, and promote green and sustainable development. The rapid growth of photovoltaic (PV) and wind turbine (WT) installations also brings many new problems. In terms of technology, intermittent renewable energy resources are characterized by strong randomness, high volatility, weak observability, and poor controllability, and large-scale access to the main grid will pose a threat to the safe and reliable operation of the main grid. In terms of operation, there are many subjects, rich types, and huge space for optimization of distributed resources, but the supporting market mechanism is not perfect and the business model is single.

Microgrid (MG) can integrate and utilize local distributed energy flexibly and efficiently, improving the reliability of local power supply and the efficiency of energy utilization [2]. MG has attracted extensive attention for its economic and environmental benefits due to its clean energy, flexible power generation mode, compatibility with the environment, and small loss. Industrial and commercial parks are important spatial clustering forms for regional economic development, industrial adjustment, and upgrading and have important functions of gathering innovative resources, cultivating emerging industries, and promoting urbanization [3]. The development of industrial and commercial parks is highly dependent on the power supply, which requires the reliability and quality of power supply. The park MG, as a highly reliable power supply technology close to source and load, interacts with large power grid and users in a friendly manner and is becoming an important location for new energy formats such as electric vehicles and multi-energy complementary and energy Internet. The new power reform encourages social capital to enter into the power distribution business including MG construction and operation, which stimulates the market vitality of MG investment and operation, making it an inevitable trend for multiple entities to invest in MG. With the gradual opening of the electricity market and the decline of subsidies for renewable energy [4], more and more distributed energy resource (DER) investors begin to seek more innovative and diversified operation modes to obtain reasonable investment returns [5]. Innovation of business mode, realization of extensive user interaction, and full exploitation of user flexibility are the important directions of park MG development.

5.1.2 Literature Review

In most of the relevant papers, the economic optimization operation of park MG in a market-oriented environment mainly focuses on operation mode, transaction mechanism and economic dispatch, etc. In terms of market operation mode, ref. [6] constructed a multi-objective optimization model of MG operation from three aspects of operating cost, pollutant emission, and operational risk, which effectively improved the multi-objective comprehensive management level of MG operation. In ref. [7], considering the operation characteristics of MG's dual role in the market environment, an optimization model of MG's direct grid-connected electricity purchase and sale was established. Ref. [8] established comprehensive benefit evaluation models of MG projects under different business operation modes and considered the impact of initial investment, operation and maintenance, investment return, and other factors on the economy of MG projects from the perspective of the whole life cycle. This model provides an accurate and effective method for the demonstration and evaluation of MG project in the early stage. In ref. [9], an index evaluation model for MG energy system was established, considering four influencing factors, including economy, reliability, energy consumption, and environmental protection.

In terms of transaction mechanism, ref. [10] proposed an MG power market transaction theory. Ref. [11] establishes a buyer-seller stochastic matching transaction model for the electricity transaction of multiple operating entities in park MG and solves the transaction problem when supply and demand are unbalanced by introducing the concept of "virtual unit." However, such kind of random matching trading mechanism has considerable randomness and does not consider the constraints of actual calculation force and time. In ref. [12], an MG optimal operation method based on multi-agent systems (MAS) was proposed to allocate the interests of DER investment subjects through competitive equilibrium mechanism. However, the bidding process did not consider the impact of the uncertainty of renewable energy power generation, and the investment subject involved one single type of DERs, without considering the role of energy storage in MG. Ref. [13] proposed a power trading mechanism with monopoly characteristics, including day-ahead scheduling and real-time scheduling. To limit the monopolistic behavior of some entities, the punishment mechanism was set up in the transaction link to ensure the settlement price is within a reasonable range. Ref. [14] proposed a grid-connected MG system with multi-time scale trading mechanism and a trading strategy optimization algorithm based on deep learning, A daily and intraday trading mechanism was also established, forming a day-ahead trading plan and clearing after the trading price. In real-time trading, MG only declares one unbalanced power demand to be cleared. Ref. [15] proposed a distributed computing and power trading framework based on blockchain intelligent contract and described the data center architecture and computing model for PV equipment. Aiming at the problem that it is difficult for PV to effectively match the energy consumption demand of data center, a new energy trading method based on blockchain is proposed to optimize the overall PV energy utilization rate.

In terms of economic dispatch, ref. [16] established a park MG two-layer dispatch model including DR layer and MG layer. DR layer optimized the load curve by means of source-charge interaction, and MG layer minimized the comprehensive operation cost based on real-time electricity price. Ref. [17] proposed the bidding game method of MG in the power market, studied the economic operation results of MG under the two strategies of alliance and non-alignment, and modelled it into a two-layer model for solution. Ref. [18] takes an industrial park as the research object and studies the energy coordination and optimization scheduling method of park MG to minimize the power supply cost.

Most of the abovementioned studies on MG operation use a single investment or operation entity as the default premise and mainly focus on the output planning of DERs, energy storage charging and discharging strategies, energy trading strategies with main grid, and controllable load dispatching to realize the economic operation of MG, but these methods are difficult to apply in multi-stakeholder situations. The adjustment of the renewable energy subsidy policy and the gradual opening of the electricity market have prompted MG to seek a variety of operating modes and mechanisms. Policies and market environment in park MG are loose, making it the front line of operating mode mechanism innovation. Research in recent years has proposed various models and mechanisms, which can achieve good results under the set market environment and provide a useful reference and basis for the commercial operation of the park MG. However, there are still problems of insufficient consideration of the operation mode, transaction mechanism, and optimized operation. For example, some studies have simplified the background of the operation mode of the transaction mechanism, making the transaction model too ideal, ignoring the calculation ability of the transaction unit and the real-time requirements of the transaction [19]. Therefore, considering the uncertainty to establish a fair and effective operating mode and benefit distribution mechanism and to use fast and efficient solving algorithms to improve the performance and stability of the MG system based on bidding trading operations, the problem needs to be solved urgently.

The artificial immune system (AIS) is an adaptive system, inspired by theoretical immunology and observed immune functions, principles, and models [20]. Since AIS has the advantages of high parallelism, distributed operation, strong adaptability, and self-organization, AIS has been implemented in various fields including computer network security, power system reconfiguration, intelligent control, pattern recognition, and failure detection [21, 22]. Ref. [23] presents a methodology for the reconfiguration of radial electrical distribution systems based on the bio-inspired metaheuristic artificial immune algorithm to minimize energy losses. A hybrid-immune-system-based particle swarm optimization is used in [24] to minimize the fuel cost for generation assuming realistic market prices for power, distributed generator bids reflecting realistic operational costs, and load bids customized according to the consumers' priorities.

5.1.3 *Contributions and Significance*

In order to realize the flexible economic operation of the park MG under the market environment and the fair distribution of internal multi-subject interests, this chapter proposes a multi-agent bidding optimization operation method based on AIS for the park MG, including the lower-level bidding process and the upper-level liquidation process. The upper level establishes a mixed-integer programming model with the goal of the lowest electricity price in the MG or the least unbalanced electricity; the lower level aims at optimizing the maximum benefits of each DER and integrates cost information, forecast information, and price information. And the results of the previous round of bidding are transformed into antigens, and the optimal bidding strategy is solved by AIS. Compared with existing methods, the methods proposed in this chapter have the following contributions:

1. Establish an internal bidding transaction mode for the park MG, and build a simple and feasible "micro market" mechanism within the multi-agent park MG to realize the fair distribution of the benefits of all parties based on ensuring the economic and safe operation of the system. The deviation cost coefficient is introduced to indicate the impact of the uncertainty of renewable energy generation on the bidding process, and the internal electricity price formed can reflect the internal supply and demand relationship of MG. As a controllable characteristic under the market environment, MG is externally manifested and uses the peak-to-valley price difference of the large power grid to increase revenue. By achieving coordination and interaction with external power grids, it supports the consumption of renewable energy, which improves energy utilization efficiency and at the same time brings social benefits of peak shaving and valley filling for main grids.
2. In order to support each subject to participate in bidding transactions, an AIS-based bidding strategy search method is proposed, which can search for highly adaptable antibodies in the unknown and drastically changing environment and electricity price information. Through the secondary immunization of AIS, it can respond quickly to antigens with similar environmental and market information. The evolution of the antibody gene library in the AIS can accelerate the maturity of antibody solutions and can better meet the requirements of real-time calculation.

5.2 Characteristic and Market Positioning of Park MG

5.2.1 *Positioning of Park MG in Market Environment*

MGs of industrial and commercial park, which have a high demand for electricity consumption and power quality, will be connected to the bulk power grid in the face of diversified social subjects and investment and realize intelligent operation and management through PIoT. Park MG with multi-agent investment will become an important form in the future, mainly for the following reasons:

1. With the separation of distribution and sales process and the development of power market, the investment threshold is lowered, the market vitality is enhanced, and the operation mode of multi-agent investment diversification is bound to appear.
2. MG with multi-stakeholder can share risks and benefits. Modular, specialized investment and full responsibility for self-profit mechanisms can improve the quality and efficiency of the construction and operation of MG.
3. PIoT technology enhances the function of MG for DERs to plug and play. It can support DERs built by power users to achieve flexible access and operation and provides technical support for multi-principal investment.

Under the background of separating power distribution and sale process, park MG is faced with new opportunities and challenges [25]. The opportunity is that social asset enterprises can apply for and invest in MGs and carry out electricity sales business and additional value-added services, lowering the market threshold and stimulating the vitality of the social capital construction of MG. The challenge is that the MG, as a new market player, needs to constantly explore and adapt to new market mechanisms and operating modes and bear the market risk caused by supply and demand fluctuations alone.

As a micro, autonomous, and friendly energy system, MG participating in the operation of the electricity sale will greatly affect the traditional power market structure. The mode of typical MG participating in the market-oriented operation is shown in Fig. 5.1. New power trading relationships will be formed between MG and grid companies, power generators, electricity retailers, and other MGs, enabling the exchange of energy, information, and services. The park MG, which acts both as a user and as a small power producer, participates in power trading and has a dual identity that is different from traditional single-line sales. The MG can sign a long-

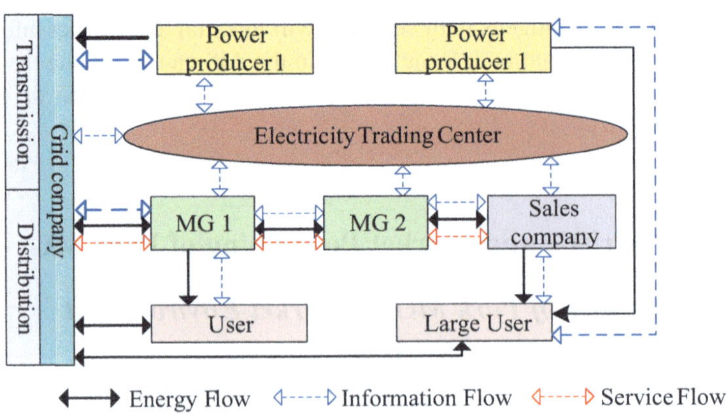

Fig. 5.1 External interaction of MG

term ancillary service bilateral contract with the grid company as an independent ancillary service provider or participate in the ancillary service transaction market and obtain ancillary service income. The MG can also sign long-term energy transaction contracts with other MGs and form alliances or multiple MG system energy transactions when participating in the power market. MG may also participate in energy transaction contracts with the electricity selling company or operate the electricity selling company under its custody in order to avoid market risks caused by small scale. As a controllable and friendly unit, the MG obtains the access service of the grid company and simultaneously reports the necessary operational information with the main grid dispatching center. It can also participate in the demand response to peak and valley electricity prices.

5.2.2 The Composition and Multi-agent Structure of MG

The concept of MAS was firstly put forward in 1986 by Marvin L. Minsky, a famous scholar in the field of distributed artificial intelligence: an interactive and interrelated system composed of multiple agents with certain communication ability and logical computing ability, which can realize the goal of the whole system through information sharing and cooperation among these agents [26]. An agent can be a hardware or a packaged computing software system, which can be applied in both the actual system and the virtual network environment to realize the interaction between multiple individuals and intelligent decision-making [27]. The basic structure of agent includes information processing module, target module, perception module, communication module, and execution module. The target module makes the optimization or working target of the agent. The perception module is responsible for environmental information acquisition and feeds back to the information processing module. Information processing module makes decisions based on existing information and data. Communication module is responsible for the communication mode and interaction information between agents. The execution module is used to complete various behaviors required by the agent and influence the environment.

According to the composition structure of MAS-based MG, each DER in MG is regarded as an agent unit that can conduct information exchange and calculation. MG is connected to the bulk grid through contact nodes, purchases power from the large power grid at price of π^{buy}, and sells power to the bulk grid at price of π^{sell}. The DER bidding-based park MG optimized operation structure mainly involves three types of agent: optimized control agent (OC agent), operation management agent (OM agent), and bidding unit agent (BU agent), as shown in Fig. 5.2.

1. OC agent: This agent sends the current operating status of MG to the OM agent and receives the power generation and load energy balance information returned

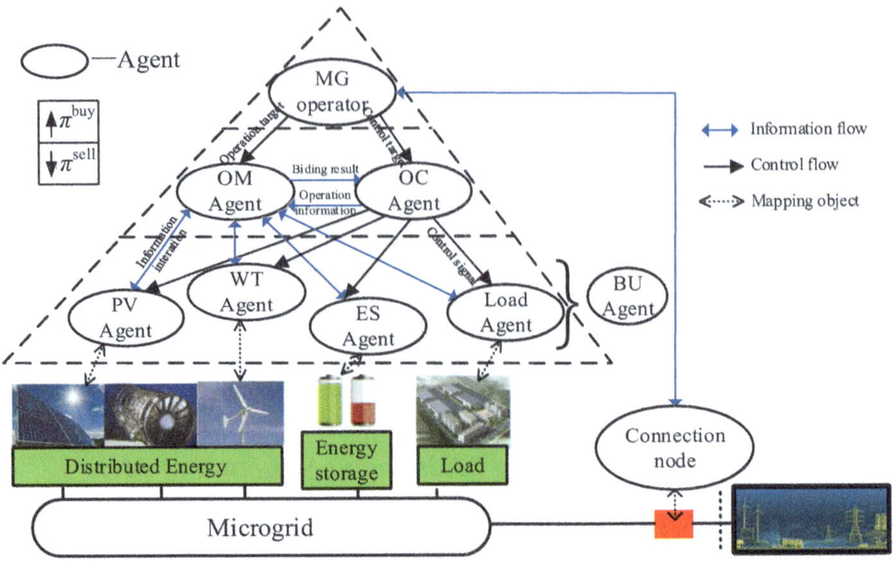

Fig. 5.2 Structure of park MG based on MAS

by the OM agent. In addition to the MG energy management, the OC agent also monitors the voltage, power quality, and frequency of MG. Under special circumstances (such as when a large grid accident occurs, MG will be transformed from grid-connected to islanded operation), it can directly control the load and DER power output within MG beyond the OM agent.

2. OM agent: This agent can assist the OC agent to complete the economic and power generation dispatch in the MG, transmit the purchase/sale price to the loads and DER unit agents, and manage the BU agents in the MG. Its main functions include obtaining each DER bid, judging whether the negotiation is completed, filing bidding information, and obtaining unit operation plans.
3. BU agent: It mainly includes energy source units used to provide power to MG, such as PV, WT, fuel cell, micro-turbine (MT), and other typical distributed energy sources. Based on the local measurement information and the communication information with other agents, the DER BU agent can submit a power generation bidding request to the OM agent. The load BU agent mainly submits the expected amount of electricity for various loads in the MG but does not participate in the bidding. The energy storage BU agent is responsible for the energy storage system in the MG and performs energy management of the energy storage system, including charge and discharge time management, charge and discharge power management, and electric vehicle charge and discharge management, and gains benefits through the formulation of charge and discharge strategies.

5.2.3 MG Bidding Feedback and Co-evolution

Bidding feedback and coevolution schemes in the bidding process of MG are shown in Fig. 5.3. The key to the bidding algorithm of DERs based on AIS is the adaptive immune algorithm of individual agents and the realization of the co-evolution of agents realized on this basis. The technical idea of MG bidding feedback and co-evolution scheme is: regard the BU agent of DER as an agent with perceptual ability and rational thinking. Agents interact through the submitted bidding information (antibody representative). The MG realizes the synergy between source and load through the bidding feedback provided by the OM agent.

An artificial immune solution process of BU agent is a bidding process. In a bidding period, the BU agent first uses its own environment and main network electricity price information as the antigen to perform artificial immune algorithm to obtain the antibody solution. After that, the OM agent collects bidding information and publishes it to each BU agent in the form of bidding feedback. The BU agent generates a new antigen based on the feedback information and recalculates the bidding strategy. After several rounds of feedback and negotiation, the final quotation is obtained.

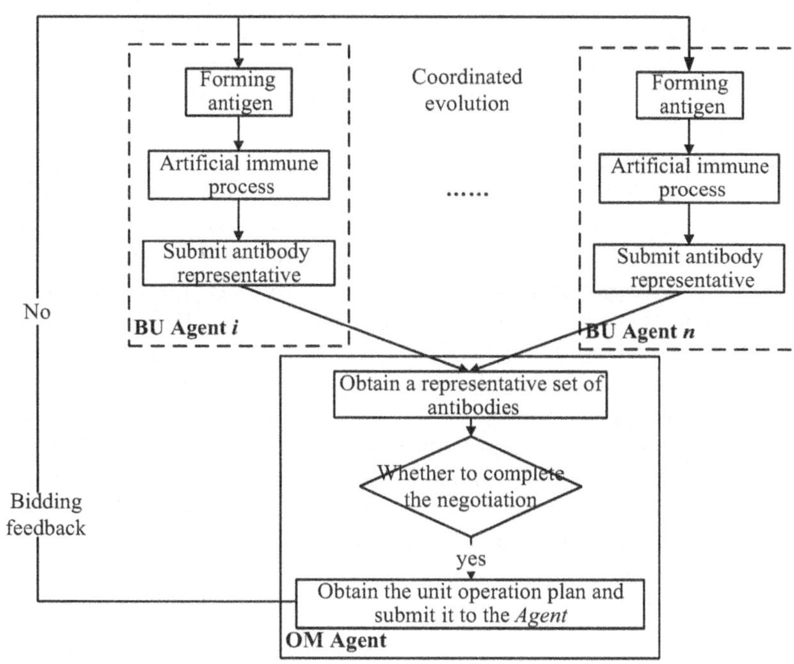

Fig. 5.3 MG bidding feedback and co-evolution

5.3 Bi-level Optimization Model for Bidding Operation

5.3.1 Multi-agent Bidding Transaction Operation Mode

At present, enterprises investing in renewable energy such as PVs mainly adopt the operation mode of "energy self-balance and surplus up-to-grid." The surplus or insufficient electricity is accepted or provided by the external power grid, which lacks management of the randomness and volatility of renewable energy generation. Under the trend of PV parity access to the main grid, considering the transmission of the power flow of the MG in the park, the grid company takes away the difference between the production and sales of renewable energy but neglects management, and renewable energy investors get little return on their investment. To achieve the efficient and fair operation of the multi-entity park MG, a double-layer bidding model based on bi-level optimization is proposed as Fig. 5.3 shows.

In the multi-agent system of the MG, the OM agent is responsible for the access and removal of the load and DERs, formulating the trading rules within the network, organizing and managing the electricity trading, forecasting the load, and managing purchase and sale in power market. The BU agent is responsible for the distributed power supply owned by the investment entity to participate in the bidding transaction decision, the power generation prediction of DERs, and the local information management of the power source. The ES system agent performs energy management of ES system, including charge and discharge times and power management, and EV charge and discharge management. When bidding for a certain period, the MG operation management agent initiates the transaction according to the load forecasting situation in the network.

The OM agent collects the bidding strategies of each BU agent and calculates the clearing price and power generation allocation of the current round and distributes it to each BU agent as the next round of bidding decision. When bidding, the BU agent updates the environment information and benefit function. The optimization strategy embedded in the BU agent can search for the bidding strategy that maximizes its own revenue. After several rounds of interaction between the BU agent and the OM agent, when each BU agent no longer changes the bidding strategy or reaches the number of bidding round threshold, the final clearing price, power generation plan, and unbalance amount are formed.

The bidding transaction mechanism of the optimal operation of MG provides a reasonable distribution model for multi-agent investment returns of MG, which realizes the balance of the autonomy and externally controllable of MG, and increases revenue through unbalanced electricity trading.

5.3.2 Lower-Level Optimization Model

The lower-level optimization model of MG bidding transaction includes BU agent model and load elasticity model. The BU agent model solves the bidding strategy with the maximum revenue of distributed power source as the target; the load elasticity model makes the load response based on each round of electricity price.

Assume that there are s DERs in the MG, corresponding to s BU agents; the bidding function of BU agent i is expressed as follows:

$$\rho_{i,t} = a_{i,t} + b_{i,t} \times P_{i,t} \tag{5.1}$$

where $a_{i,\ t}$ and $b_{i,\ t}$ are the price coefficients of the bidding function that DER i submitted to the OM agent. $P_{i,\ t}$ is the amount of generated energy with unit kWh, and $\rho_{i,\ t}$ is the bidding price of generation $P_{i,\ t}$ under the bidding strategy $a_{i,\ t}$ and $b_{i,\ t}$ with unit CNY/kWh.

The MG exchanges power with the external grid through the tie line, and the tie line can also be regarded as the bidding unit participating in the price clearing in the MG. The price of electricity purchased by the MG through the tie line is λ_t^{buy}, and the price of electricity sold through the tie line is λ_t^{sell}. The bidding function of the tie line can be expressed as

$$\lambda_t^{\text{grid}} = 0.5 \times \left[\lambda_t^{\text{buy}}\left(1 + \text{sgn}\left(P_t^{\text{tie}}\right)\right) + \lambda_t^{\text{sell}}\left(-1 + \text{sgn}\left(P_t^{\text{tie}}\right)\right)\right] \tag{5.2}$$

where λ_t^{grid} represents the price of electricity exchanged through the tie line; when P_t^{tie} is positive, it represents the price of electricity purchased; and when P_t^{tie} is negative, it represents the price of electricity sold. P_t^{tie} is the energy transmitted by the tie line. Sgn $(\cdot)$ stands for symbolic function.

Based on the new round of electricity price and power generation allocation plan issued by the MG operation center, the BU agent i determines the price coefficient of the next round of bidding with the goal of maximizing its own revenue. In the nth round of bidding, the objective function of the BU agent i can be written as

$$\max B_{i,t,n}(a_{i,t,n}, b_{i,t,n}) = P_{i,t}\lambda_{i,t,n}^{\text{cl}} - C_{i,t,n}(P_{i,t}) \tag{5.3}$$

where $\lambda_{i,t,n}^{\text{cl}}$ is the clearing price of electricity in the MG after the nth round of bid, solved by the upper-level models (5.9)–(5.17).$C_{i,\ t,\ n}(P_{i,\ t})$ is the cost function of BU agent i, expressed as

$$C_{i,t,n}(P_{i,t}) = c_{i,1}P_{i,t}^2 + (c_{i,0} + \sigma_{i,t})P_{i,t} + c_{i,\text{const}} + U_{i,t}C_{i,\text{startup}} \tag{5.4}$$

where $c_{i,\ 1}$ and $c_{i,\ 0}$ are the cost coefficients of DER i. $c_{i,\text{const}}$ and $C_{i,\text{startup}}$ are the fixed cost and startup cost, respectively. It should be noted that the operating cost of renewable energy like PV and WT is 0. $U_{i,t}$ is a binary variable, and $U_{i,t} = 1$ when

DER i starts up; otherwise, $U_{i,t} = 0$. Considering the impact of deviation penalty on cost, $\sigma_{i,\ t}$ is the penalty coefficient of predictive deviation, expressed as

$$\sigma_{i,t} = \sum_{t \in \mathrm{T}} \left| P_{i,t}^{*} - P_{i,t} \right| / P_{i,t} \tag{5.5}$$

where $\boldsymbol{T}$ is a set of time periods that have similar characteristics to the current environment, such as the same period of the previous few days, or a period with similar weather prediction values; $P_{i,\ t}$ is the bidding generation quantity of t period; and $P_{i,t}^{*}$ is the actual generation quantity of t period. At the same time, the output of the distributed power supply is also subject to the upper and lower limits of its own power generation and the climbing slope constraints:

$$P_{i,t,\min} \le P_{i,t} \le P_{i,t,\max} \tag{5.6}$$

$$-R_i^D \le P_{i,t} - P_{i,t-1} \le R_i^U \tag{5.7}$$

where $P_{i,\ t,\ \min}$ and $P_{i,\ t,\ \max}$ are the upper and lower limits of the output of the distributed power supply, respectively. R_i^D and R_i^U are the upward and downward climb rate limits of the DERs in unit time Δt.

The types of load can be divided into three categories according to their different electrical characteristics, i.e., uncontrollable loads, transferable loads, and variable loads. This chapter only considers the uncontrollable load and the variable load. Combined with the typical demand elasticity curve in economics, the load model can be simplified to the load elasticity curve. The load demand at the nth bid is as follows:

$$D_{t,n}^{\text{load}} = \sum_{j=1}^{N_{\mathrm{C}}} D_{j,t} + \sum_{h=1}^{N_{\mathrm{H}}} \left(\alpha_h \lambda_{t,n-1} + \beta_h \right) \tag{5.8}$$

where C_j represents the jth uncontrollable load and α_h and β_h represent the response coefficients of the hth load. N_{C} is the number of uncontrollable loads, and N_{H} is the number of variable loads.

In the case of high uncertainty, the load prediction error is large. Equation (5.8) should be replaced by a stochastic model predictive control (stochastic MPC) method which can be found in [28, 29].

5.3.3 Upper-Level Optimization Model

The upper-level optimization model of MG bidding transaction includes OM agent model and ES agent model. The P_{cl}^{n} is optimized by the OM agent based on all the submitted bidding functions with nonlinear constraints, and the price ρ and the power distribution allocation P are the optimization variables. The calculation formula can be transformed into a mixed-integer linear programming mode, expressed by formulas (5.9)–(5.17). The optimization variables ρ and P are replaced by new continuous optimization variables $\boldsymbol{w}$ and binary variables z.

$$\min k_1(d^+ - d^-) + k_2\lambda_{t,n}^{\mathrm{cl}} \tag{5.9}$$

$$P_{t,n}^{\mathrm{tie}} + d_{t,n}^{+} - d_{t,n}^{-} = P_t^{\mathrm{tie\_}tg} + D_{t,n}^{\mathrm{load}} - \sum_{i=1}^{s} P_{i,t,n} \tag{5.10}$$

where $P_{t,n}^{\mathrm{tie}}$ is the unbalanced power of the nth bidding round. $d_{t,n}^{+}$ and $d_{t,n}^{-}$ are the positive and negative deviation of the unbalanced power target value, respectively, and non-negative values. k_1 and k_2 are the objective weight adjustment coefficients, which are changed according to the specific operating state and task. $\lambda_{t,\ n}$ is the clearing price of the park MG in the nth round. $P_t^{\mathrm{tie\_}tg}$ is the tie line control target when the park MG participates in the cloud application like transaction or demand response and is zero when MG is in autonomous state. The constraints that the model needs to satisfy at the same time are Eqs. (5.6)–(5.8) and (5.11)–(5.17).

$$\lambda_{t,n}^{\mathrm{cl}} = \rho_{1,t,n} = \rho_{2,t,n}\cdots = \rho_{1,t,n} \tag{5.11}$$

Among them, formula (5.11) is the market unified clearing price constraint; that is, the electricity price of all DERs is consistent.

$$P_{i,t} = w_{i,1}P_{i,t,\min} + w_{i,2}P_{i,t,\mathrm{lower}} + w_{i,3}P_{i,t,\mathrm{upper}} + w_{i,4}P_{i,t,\max}, \forall i = 1,\cdots,s \tag{5.12}$$

$$\rho_{i,t} = w_{i,1}\rho_{i,t,\min} + w_{i,2}\lambda_t^{\mathrm{sell}} + w_{i,3}\lambda_t^{\mathrm{buy}} + w_{i,4}\rho_{i,t,\max}, \forall i = 1,\cdots,s \tag{5.13}$$

$$\lambda_t^{\mathrm{sell}} \le \rho_{i,t,n} \le \lambda_t^{\mathrm{buy}},\ \forall i = 1,\cdots,s \tag{5.14}$$

$$0 \le w_{i,1} \le z_{i,1}, 0 \le w_{i,2} \le z_{i,1} + z_{i,2}, 0 \le w_{i,3} \le z_{i,2} + z_{i,3}, 0 \le w_{i,4} \le z_{i,3},\ \forall i = 1,\cdots,s \tag{5.15}$$

$$w_{i,1} + w_{i,2} + w_{i,3} + w_{i,4} = 1, \forall i = 1,\cdots,s \tag{5.16}$$

$$z_{i,1} + z_{i,2} + z_{i,3} = 1,\quad z_{i,1},\ z_{i,2},\ z_{i,3} \in \{0,1\},\ \forall i = 1,\cdots,s \tag{5.17}$$

where s is the number of DERs that participates in the MG bidding, D_{load} is the load required at this period, $P_{i,\ t,\ \min}$ and $P_{i,\ t,\ \max}$ are the generation boundaries of DER i, $\rho_{i,\ t,\ \min}$ and $\rho_{i,\ t,\ \max}$ are the prices at which DER bids at the generation boundaries, λ_t^{buy} and $\lambda_t^{\mathrm{sell}}$ are the market price range for electricity purchase and sale of DERs, and $P_{i,\ t,\ \mathrm{lower}}$ and $P_{i,\ t,\ \mathrm{upper}}$ are the bid generation when the market electricity price is at the boundaries. w and z are the variables generated after linearizing the bidding piecewise function. $w$and $z$are the continuous variable and the 0–1 control variable.

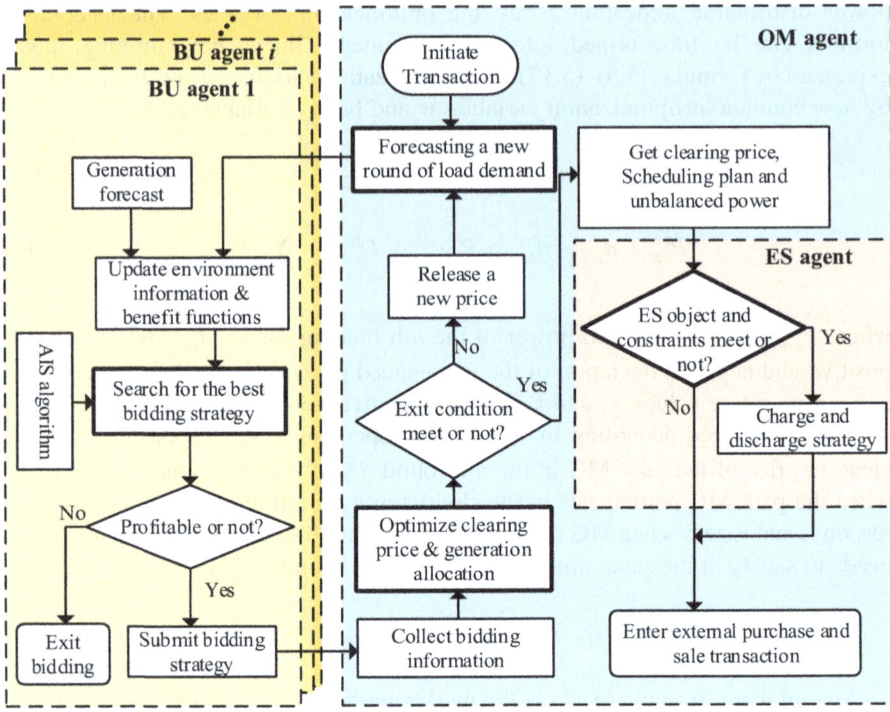

Fig. 5.4 Multi-agent MG bidding transaction mechanism process

Figure 5.4 shows four different bidding functions. $P_{i,\ \min}$ and $P_{i,\ \max}$ are the generation boundaries of DER i, $\rho_{i,\ \min}$ and $\rho_{i,\ \max}$ are the prices at which DER bids at the generation boundaries, $\rho_{i,\ \text{lower}}$ and $\rho_{i,\ \text{upper}}$ are the market price range for electricity purchase and sale, and $P_{i,\ \text{lower}}$ and $P_{i,\ \text{upper}}$ are the bid generation when the market electricity price is at the boundaries. For BU agent 1, there are $\rho_{\min} = \rho_{\text{lower}}$ and $\rho_{\max} = \rho_{\text{upper}}$. For BU agent 2, there are $P_{\min} = P_{\text{lower}}$ and $P_{\max} = P_{\text{upper}}$. For BU agent 3, there are $\rho_{\max} = \rho_{\text{upper}}$ and $P_{\min} = P_{\text{lower}}$. For BU agent 4, there are $\rho_{\min} = \rho_{\text{lower}}$ and $P_{\max} = P_{\text{upper}}$.

According to the unbalanced power supply and future electricity price interval information during this period, the ES system agent will judge whether the unbalanced power can meet its charge and discharge power constraints, capacity constraints, and economic constraints and develop a charge and discharge plan for maximizing the daily benefit of ES. The objective function is as follows:

$$\min \sum\nolimits_{t=1}^{24} P_t^{\text{B}} \lambda_t \Delta t \tag{5.18}$$

subjected to

$$\max\left\{-P^{\text{es}}_{\text{dis\_max}}, \frac{\Delta P_{\text{unblc}} - |\Delta P_{\text{unblc}}|}{2}\right\} \leq P^{\text{B}}_t$$
$$\leq \min\left\{P^{\text{es}}_{\text{ch\_max}}, \frac{\Delta P_{\text{unblc}} + |\Delta P_{\text{unblc}}|}{2}\right\} \tag{5.19}$$

$$E(t) = E(t-1) + P^{\text{B}}_t \Delta t \tag{5.20}$$

$$E^{\text{es}}_{\text{min}} \leq E(t) \leq E^{\text{es}}_{\text{max}} \tag{5.21}$$

$$E(24) = E(0) \tag{5.22}$$

where, P^{B}_t is the charge and discharge power of ES at time t, which represents the charging power when it is positive. λ_t represents the price of electricity at time t; ΔP_{unblc} is the unbalanced power of the MG; $-P^{\text{es}}_{\text{dis\_max}}$ and $P^{\text{es}}_{\text{ch\_max}}$ are the upper and lower limits of the charge/discharge power per unit time of the ES device, respectively. $E(t)$ is the capacity of the ES device at time t. $E^{\text{es}}_{\text{min}}$ and $E^{\text{es}}_{\text{max}}$ are the upper and lower limits of the capacity of the ES device. In the daily optimization scheduling, the ES should also satisfy the ES cycle constraints (5.22).

5.4 DER Bidding Process Based on AIS

In order to solve the model of the BU agent in Sect. 4.3, so that it can quickly obtain the bidding strategy that maximizes its own profit under the fluctuating and changing environmental information or market information, this section applies AIS to solve the bidding strategy. AIS is a type of computing system based on the function, principle, basic characteristics of the biological immune system, and related theories of immune theory to solve various complex problems [30]. Because the advantages of high parallelism, distributed operation, strong adaptability, self-organization, etc., AIS is used in many fields such as computer network security, power system reconstruction, intelligent control, pattern recognition, and fault detection [31].

5.4.1 BU Agent Bidding Process Based on AIS

The bidding algorithm based on AIS is to search the antibody that matches antigen best, and the flowchart of that searching process is shown in Fig. 5.5.

Firstly, the antibodies are encoded; then, the antibodies are induced to transfer randomly by affinity; last, the antibodies are decoded, and bidding strategies are obtained. The corresponding AIS antigens include (1) DER's prediction of environmental information during operation periods, such as light intensity and wind speeds; (2) main grid electricity purchase and sale price information; and (3) bidding information of other bidding units provided by the BM agent.

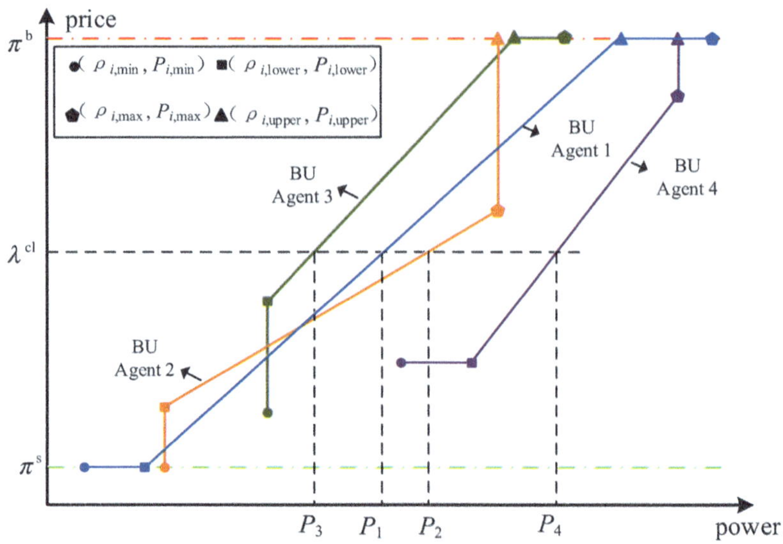

Fig. 5.5 The schematic diagram of bidding clearing

The antibodies in the first response have memory properties such that the bidding strategies can be quickly formed in the second response for similar situations. Moreover, each agent can be optimized in parallel to increase computational efficiency. The key of the bidding process is the random optimization of antibodies induced by affinity in primary response (step 5), including the selection and clone of antibodies and the setting of a corresponding gene database.

5.4.2 *Clone and Selection*

A 10-bit binary code is used to represent the decision variables a_i and b_i. The gray coding is applied to improve the local search ability of the AIS algorithm. The affinity between the antibody and antigen determines the quality of this antibody, calculated by

$$\text{Affinity}(Ab) = \begin{cases} \pi_i(a_i, b_i) + C_{\min}, & \text{if } \pi_i(a_i, b_i) + C_{\min} > 0 \\ 0, & \text{if } \pi_i(a_i, b_i) + C_{\min} \le 0 \end{cases} \tag{5.23}$$

where $C_{\min}$ is an appropriate small value. Generally, it is a negative value of fixed production cost apportioned power generation of DER i. π_i (a_i, b_i) stands for profits for the BU agent based on the bidding strategy (a_i, b_i) and power generation of DER i, calculated in (5.3).

In a generation of evolution, antibodies form a new group of antibodies by clone, mutation, and compression. To speed up the maturing of antibodies and accelerate bidding process, the scale of the clone for each antibody should be set proportional to its affinity to increase the chance of local search on the antibodies with higher affinity. The number of clones, q_i, can be expressed as follows:

$$q_i = \mathrm{Int}\left(m \cdot \mathrm{Size}_{\mathrm{clone}} \cdot \left(\mathrm{Affinity}(Ab_i) / \sum_{i=1}^{m} \mathrm{Affinity}(Ab_i) \right) \right) \tag{5.24}$$

where m is the size of antibody set, $Affinity(Ab_i)$ is the affinity value of antibody $Ab_i(t)$, and $Size_{\mathrm{clone}}$ is a parameter related to the clone scale of antibody $Ab_i(t)$. When DERs participate in the real-time market, the setting of $Size_{\mathrm{clone}}$ needs to consider the memory and calculation speed of the agent chip and the running time of MG auction.

A combination of hypermutation and receptor editing is implemented in this chapter to increase the local and global searching ability of the BU agent for getting a better bidding strategy. The application of hypermutation and receptor editing is demonstrated in Fig. 5.8. For any antibody, the mutation follows these steps:

Step 1: Set the threshold value of mutation $P \in [0, 1]$, which is normally in the range 5–15%, depending on the specific generator type of PV, WT, or MT.
Step 2: Generate a random variable p_{random} in [0, 1].
Step 3: If $p_{\mathrm{random}} > P$, the antibody mutates with hypermutation; if $p_{\mathrm{random}} \leq P$, the antibody mutates with receptor editing.
Step 4: Replace the cloned original antibodies with mutated antibodies in a generation of antibodies.

Renewable resources like PV and WT have a larger P to promise strong global search ability, considering the uncertainty of environment and volatility of power generation reflected in the antigens, while conventional resources like MT have a relatively smaller P to boost speed for a repetitive operating environment. Figure 5.6 shows the change of solution shape space for antibodies' affinity in cloning and mutation. Hypermutation makes the antibody affinity gradually climb to local optima, while receptor editing can produce new antibodies out of the current partial area to prevent prematurity and addiction to local extreme values.

5.4.3 *The Setting of Gene Database*

Antibodies are saved in an antibodies gene database. As the operation bidding in MG is periodic with high repeatability, related gene sections of most optima and similar solutions can be extracted from the memory. Gene database keeps evolving by learning from continuous feedback of the MG bidding results. In each generation of evolutions, the density of all genes attenuates by the rate ρ; then feedback is given to the corresponding gene by stimulation. If not stimulated for a long time, the

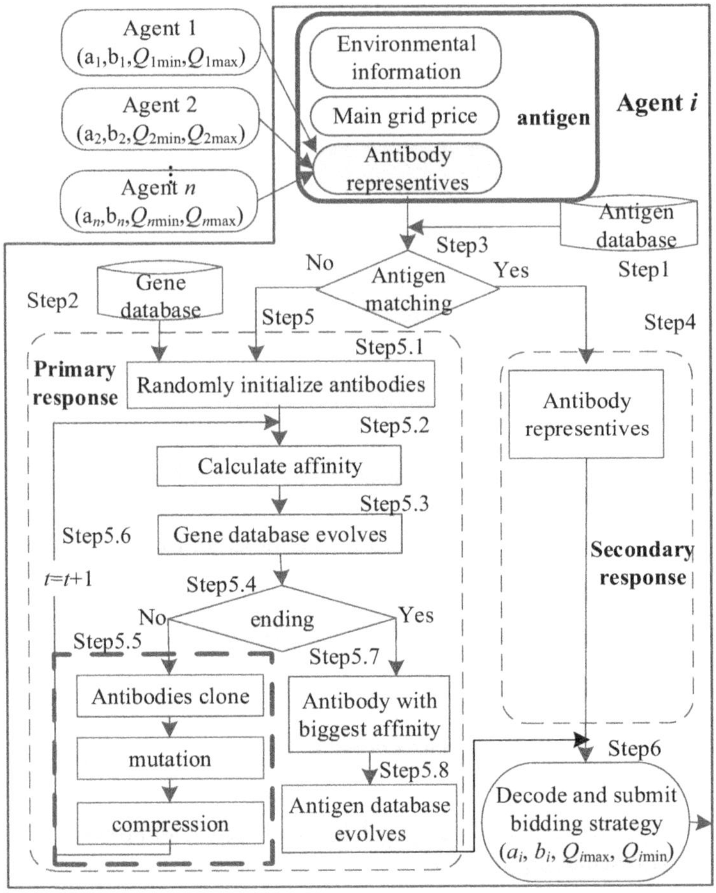

Fig. 5.6 The flowchart of the bidding process based on AIS

antibody genes will be eliminated, and those that originally highly match the antigen will be enhanced by mutation. The density of the gene database between generations is calculated by

$$
\begin{aligned}
d(x_i)' &\leftarrow (\rho + (1-\rho)\cdot \Delta d(x_i))\cdot d(x_i)\\
\Delta d(x_i) &= \begin{cases} (\text{Affinity}(Ab_{oki})/avg_{\text{Affinite}}) & \textit{if} \;\; x_i \in Ab_{oki} \\ 0 & \text{otherwise} \end{cases}
\end{aligned}
\tag{5.25}
$$

where $d(x_i)$ denotes the density of gene x_i (a_i or b_i) in gene database, ρ ($0 < \rho < 1$) is the attenuation rate, and Ab_{oki} denotes the mature antibodies i whose affinity is higher than the average value of all the antibodies ($avg_{Affinite}$). For a set with an antibody quantity of m, $avg_{Affinite}$ is calculated by

$$avg_{\text{Affinite}} = \sum_{i=1}^{m} \text{Affinity}(Ab_i)/m, \quad ab_i \in A(t) \tag{5.26}$$

5.4.4 MAS Bidding Optimization Based on Immune Co-evolution

The BU agents' search can be achieved in parallel without central control of the BM agent with AIS. Therefore, the MG's bidding process based on AIS is the co-evolution of DERs' agent antibodies and antigen. The BM agent executes feedback and coordination in the bidding process by the following steps:

Step 1: Send the antibody representatives to every BU agent.
Step 2: Wait and receive the bids submitted by all BU agents.
Step 3: Determine whether the market clearing process is accomplished. The judgment conditions consist of the following: (a) the number of feedbacks to reach its maximum; (b) the clearing price P_{cl} in the MG will not change with a certain accuracy in new feedbacks.
Step 4: If yes, record all the BU agents' bidding strategies, obtain the DER operation plan, and submit to the OM agent.
Step 5: If no, turn to step 2.

The above bidding process can quickly form a relatively better solution, gradually improve the solution quality by repeating the bidding several times, and be interrupted at any time to export a solution with a certain quality.

5.5 Case Study

5.5.1 System Parameters

The park MG to be tested is shown in Fig. 5.7. There are two PVs, two WTs, two MTs, and one ES in the MG. Four enterprises participate in the bidding process, and the configuration of them is shown in Table 5.1. There are also one energy service provider and one MG operator, maintaining the power supply and demand balance. The system voltage level is 10 kV. Due to the short distribution line of MG, the line loss has little influence on the optimization strategy. Therefore, this chapter ignores the effect of line loss in bidding and clearing. In this chapter, it is assumed that the large power grid adopts time-of-use (TOU) electricity price, as shown in Table 5.2. There are three electric prices in one day, which are peak price, flat price, and valley price, reflecting the power consumption and generation peak and valley. Figure 5.8 shows the typical daily renewable energy generation curve and load curve in three enterprises. It can be seen that the power generation of RESs is obviously insufficient at night, and the power supply will mainly rely on MTs or main grid. Operating

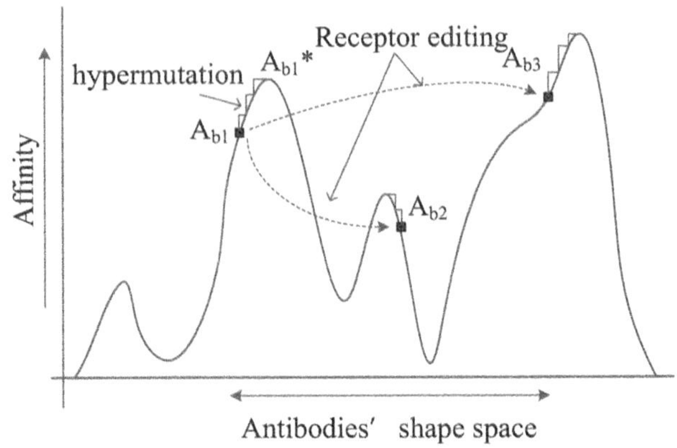

Fig. 5.7 Schematic of change of shape space for antibodies' affinity

Table 5.1 MG configuration and parameter settings

Entity	DERs	Capacity (kW)	Maximum load (kW)
Enterprise 1	PV1	3000	2500
Enterprise 2	PV2	4000	2200
Enterprise 3	–	–	3000
Enterprise 4	WT1	3000	–
	WT2	1000	–
Energy service provider	MT1	2500	–
	MT2	1500	–
MG operator	ES	1500	–

Table 5.2 Electricity price range

Time	Price attribute	λ^{buy} (CNY/kWh)	λ^{sell} (CNY/kWh)
[8, 12], [18, 21]	Peak	0.9	0.48
[13, 17], [22, 32]	Flat	0.64	0.35
[1, 7]	Valley	0.38	0.18

Table 5.3 DER parameters in MG campus

DERs	P_{min} (kW)	P_{max}(kW)	R_i^D (kW)	R_i^U (kW)	c_1(¥/kWh^2)	c_0(¥/kWh)
PV1	0	3000	–	–	0	0
PV2	0	4000	–	–	0	0
WT1	0	3000	2000	−2000	0	0
WT2	0	1500	1200	−1200	0	0
MT1	500	3000	1500	−1500	0.000315	0.7
MT2	300	2000	250	−160	0.0005	0.7

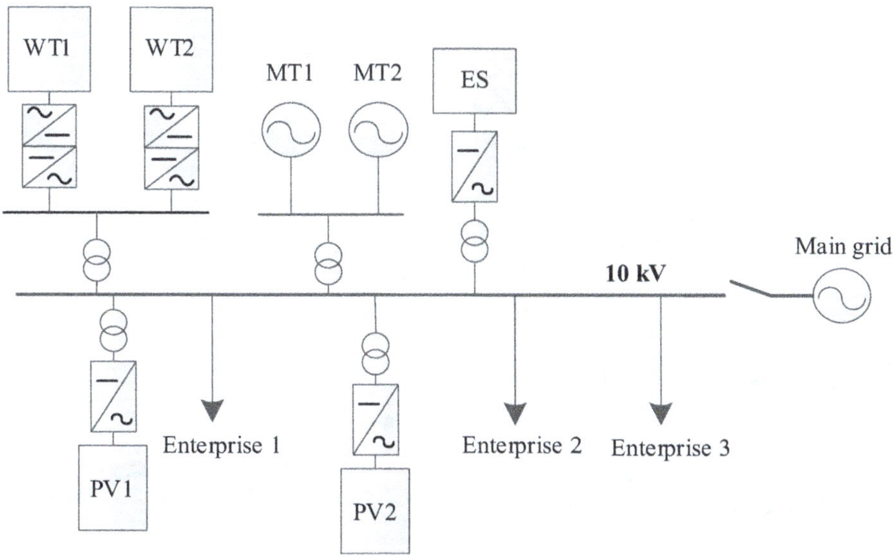

Fig. 5.8 The network structure of the tested MG

parameters of PV, WT, and MT in park MG are shown in Table 5.2 [32]. The costs of RESs are regarded as constants.

In order to verify the effectiveness of the MAS-based optimal operation of microgrid multi-agent bidding and the AIS-based optimal bidding strategy search method proposed in this chapter, this chapter takes MG multi-investor park as an example to conduct simulation analysis. The code was written in MATLAB 2018 and runs on an Intel® Core™ i5 1.99 GHz computer with 8 GB RAM (Fig. 5.9).

5.5.2 Optimal Operation Results of MG

The simulation of the MG under the optimization operation strategy proposed in this chapter was run 500 times, and each CPU time was counted with a probability of over 90% to complete the solution within 7–11 seconds, which is acceptable for the MG system. The results of the optimization are shown in Figs. 5.10 and 5.11. Figure 5.10 shows the electricity price curve in the MG under the multi-agent bidding optimization operation strategy. Figure 5.11 shows the results of DER bidding power generation, ES, and tie-line optimization.

It can be seen from Figs. 5.10 and 5.11, when the main grid is at the valley price, which means the external grid is at valley consumption, due to insufficient wind power and PV power in MG and low electricity price range outside MG, the λ_t^{cl} is at the upper limit of range, but still not high. DERs like DE decide not to participate in the market due to possible losses at low prices. The resulting power shortage will be purchased from the tie line, which, from the perspective of the main grid, promotes

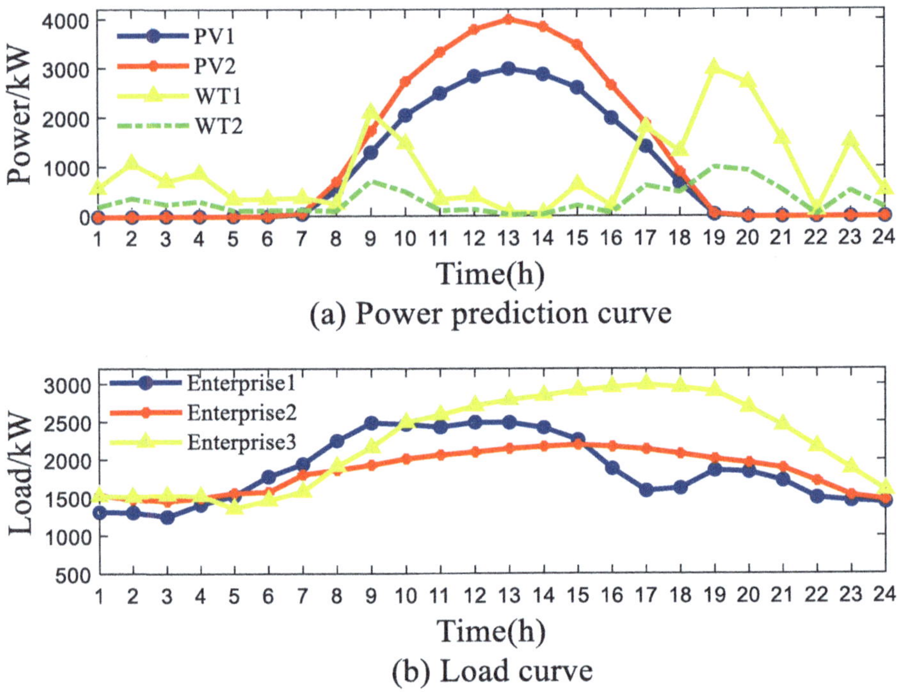

Fig. 5.9 Forecasting curves for distributed power generations and loads

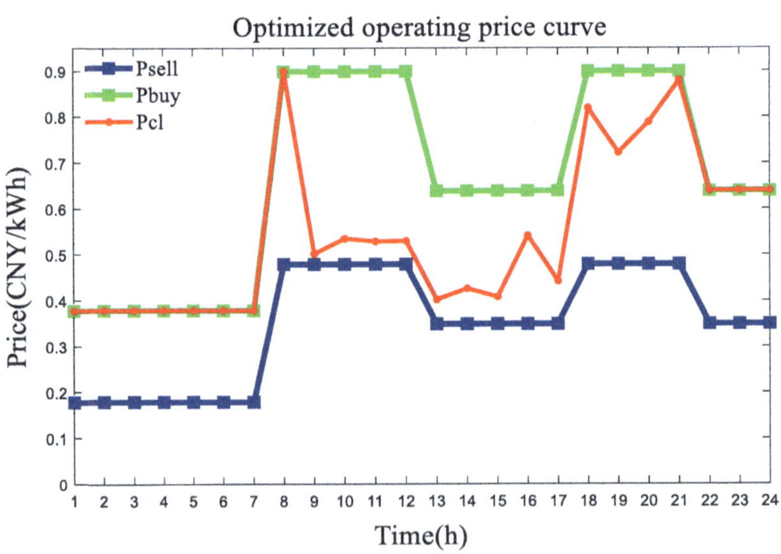

Fig. 5.10 Price curve of the optimized operating mode

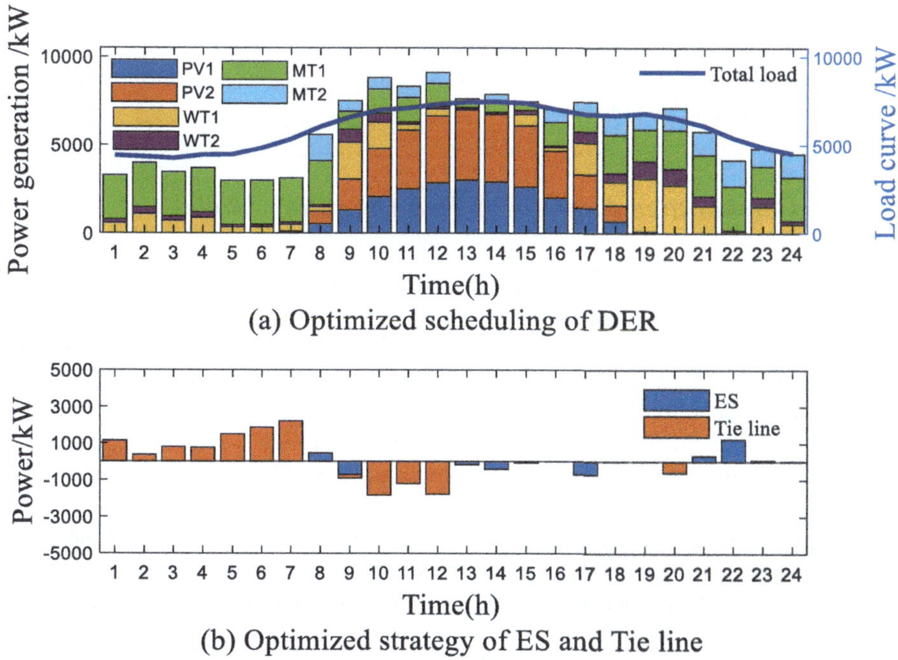

Fig. 5.11 The scheduling of (a) DERs, (b) ES, and tie line

electricity consumption of the main grid. At the first peak price, due to the high price environment and the abundant wind and PV power in MG, the bidding reduced the λ_t^{cl} to near the lower limit except 8 o'clock for a shortage of renewable energy. With high electricity prices to supplement the shortage of electricity, the ES could avoid the electricity purchase from the main grid. When the main grid is at the flat price, with price environment not so good and sufficient power generation resource, the full competition among DERs leads to an intermediate price between λ_t^{sell} and λ_t^{buy}. When at the second peak price, due to the good price environment, but reduction of the power generation, mainly from the PV, and the wind fluctuation, the λ_t^{cl} rises to the upper limit of the price range with fluctuation. When at the second flat price, due to insufficient wind power and PV power in MG, the λ_t^{cl} is at the upper limit of range, different from the valley period; the λ_t^{cl} has higher profit space for nonrenewable DERs. The ES can set the operation mode according to different operation targets and optimize the charge and discharge strategy with the goal of maximizing the full-day profit. During the valley price, due to the inadequate power supply and lower price, the MG tends to buy power from the main grid rather than increase the outputs of MTs. As the output of RES grows during the day, the outputs of MTs begin to reduce The outputs of the MTs do not reduce to zero because of their high startup cost. That MT1 was the only used generator during beginning hours is also for this reason. The redundant power will be sold to the main grid when λ_t^{sell} is high like hours 10, 11, and 12. During lower λ_t^{sell} hours like hours 14 and 17, the redundant power is stored in ES. The results show that the main grid can affect the price of MG and the power generating behavior of DERs by adjusting the price range. This is an effective way of interacting between the MG and the main grid, further enabling price-based demand response.

5.5.3 Comparative Analysis of Two Operating Modes

Under the traditional operation mode, enterprise 1, enterprise 2, and enterprise 3 adopt the strategy of self-sufficiency and surplus online. Wind company like enterprise 4 sells electricity directly to the main grid. Considering the future market environment and the reduction of renewable energy subsidies, the on-grid price of power users can only follow the price range. To verify the superiority of the proposed method, the trading price and benefit of enterprises 1–4 under the proposed mode and traditional mode are compared as shown in Figs. 5.12, 5.13, 5.14, and 5.15.

The power loads like enterprise 1 and enterprise 2 are configured with renewable energy. Taking enterprise 1 as an example, when the total trading power is positive (as shown in Fig. 5.12b period 11–16), the transaction price of the optimized model

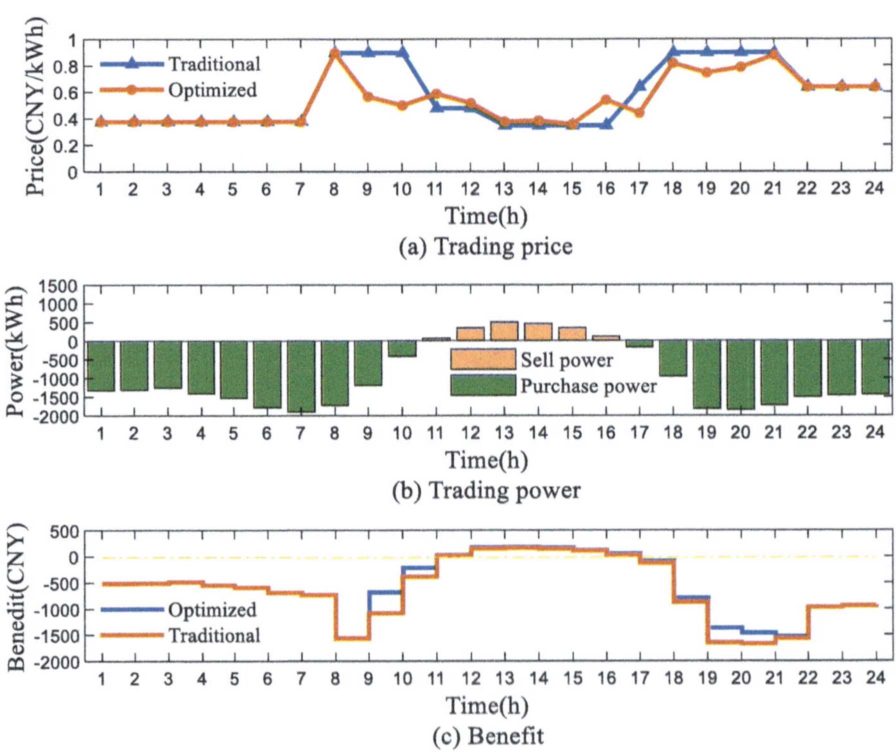

Fig. 5.12 Benefit of enterprise 1 in two modes

is higher than the price of the traditional model, shown in Fig. 5.12a. On the contrary, when the total trading power is negative, the transaction price of the optimized model is lower than that of the traditional model. This means the electricity purchase price of the optimized operation mode is not higher than the purchase price of the traditional operation mode. The electricity sales price of the optimized operation mode is not lower than the electricity sales price of the traditional operation mode. So the benefits of enterprise users in optimizing their operating models are always higher than traditional operating models as Fig. 5.12c shows. Figures 5.13, 5.14, and 5.15 present the same conclusion.

Compared with enterprise 1 and enterprise 2, although enterprise 3 does not have PV installed, it does not have revenue from PV sales, but its electricity consumption under the optimized operation mode is still lower than the traditional operation mode. Enterprise 4 earns revenue by selling the power of the WT. Figure 5.15 shows that it can sell the power to MG at a price higher than the feed-in tariff most of the time. Therefore, the proposed multi-agent MG optimal operation mode ensures the increase of the income of each investment entity while reducing the electricity price.

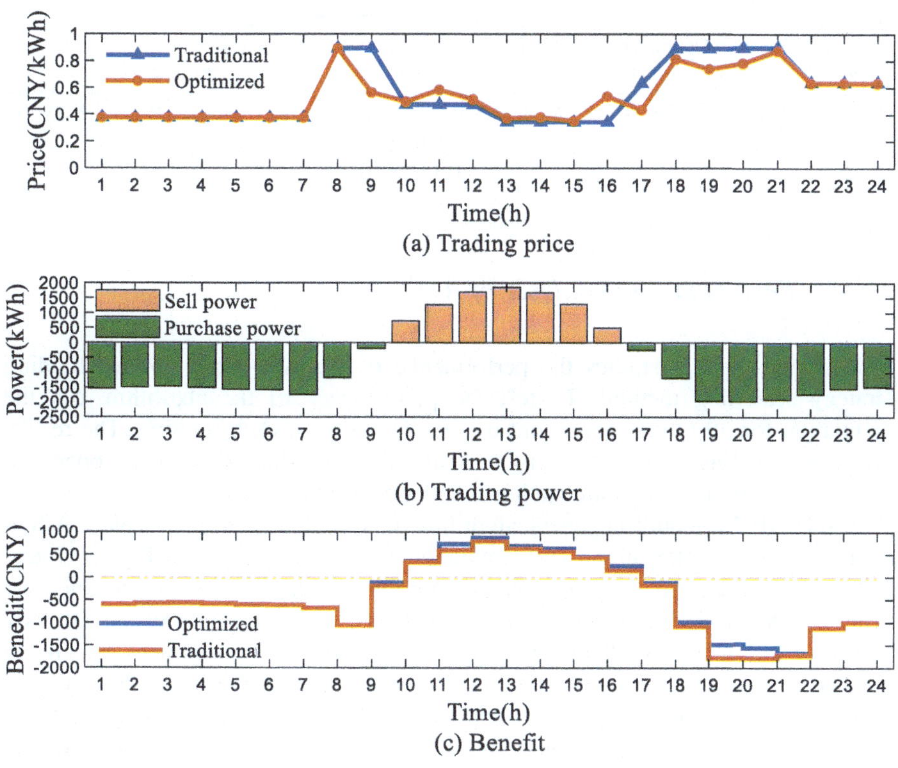

Fig. 5.13 Benefit of enterprise 2 in two modes

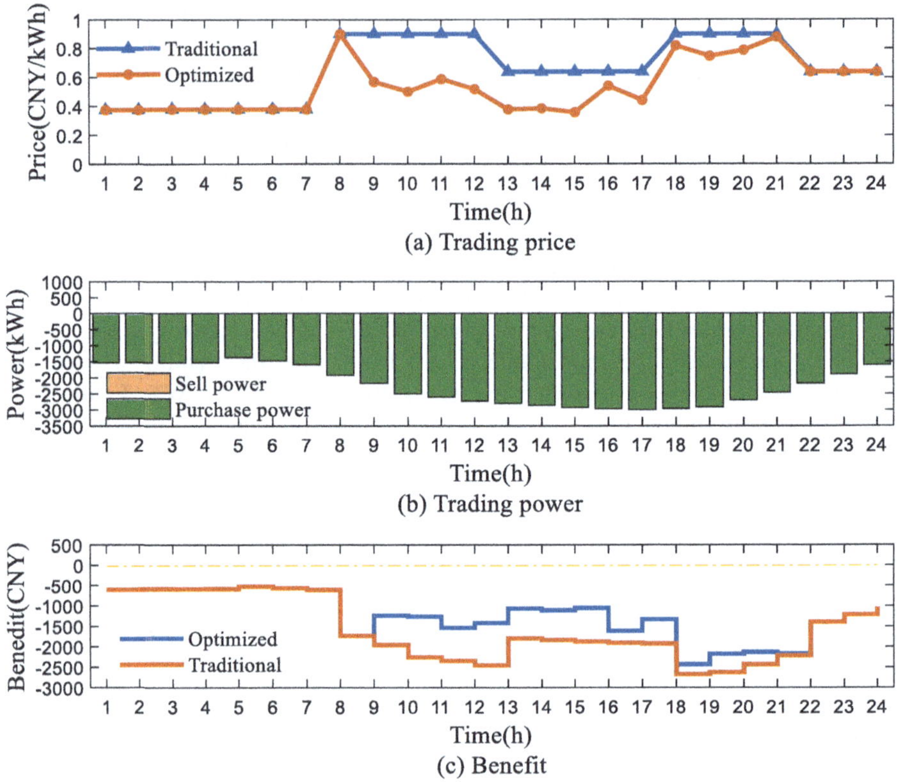

Fig. 5.14 Benefit of enterprise 3 in two modes

5.5.4 *Performance Analysis of AIS*

This section mainly studies the performance of the AIS-based optimal bidding strategy searching method. To test the performance of the algorithm, run the simulation for 500 times and count the running time of each running. The results are shown in Fig. 5.16. According to statistics, more than 90% of the operation results can complete the optimization calculation within 11 s.

MATLAB 2018 built-in genetic algorithm (GA) toolbox is used to replace AIS to achieve optimal calculation. The simulation results are shown in Fig. 5.17 and Table 5.4. AIS can be used to converge with fewer rounds and a shorter time (one order of magnitude lower than GA). This is because the evolution of the antibody gene base can retain excellent antibody genes in a similar environment, which can provide a good initial value solution for solving the problem when the antibody population is initialized, and save population size (memory) and iteration times (time) for solving the optimal antibody population. In practical application, considering the limited computing capacity of the BU agent and the real-time requirement of the system, AIS solution time is more feasible than GA.

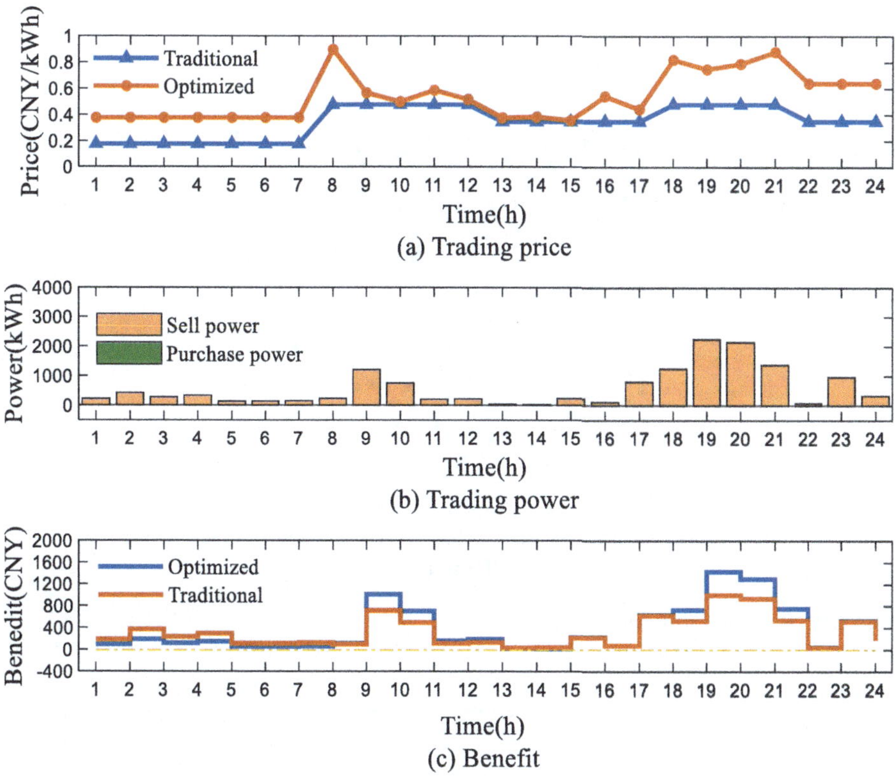

Fig. 5.15 Benefit of enterprise 4 in two modes

5.6 Conclusions

Under the background of the new power reform, in order to realize the economy, flexible operation, and fair distribution of interests of MG in multi-agent park under the market environment, this chapter puts forward a multi-agent bidding optimization operation method of MG based on AIS. In lower level, the BU agent maximizes its own revenue and the AIS to conduct the optimal bidding strategy search and submits bidding strategy to the upper operation management agent. In upper level, the OM agent obtains the settlement price and power generation distribution plan through the liquidation model and releases feedback to the lower level. The upper and lower levels interact until the settlement price converges. Finally, an example of MG in the park is given to demonstrate the effectiveness of the proposed model and method. The main conclusions are as follows:

1. The simulation results in 1.5.2 show that the proposed model and method can effectively realize balanced operation based on DER bidding within MG and

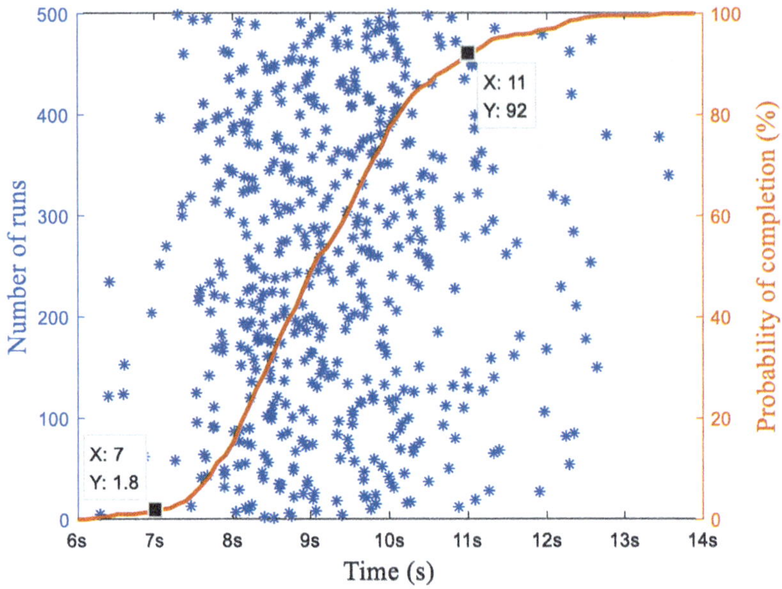

Fig. 5.16 Run 500 performance tests

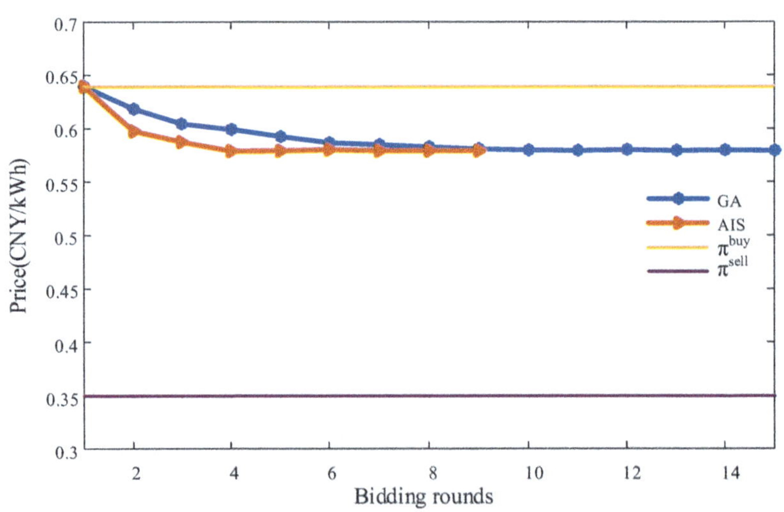

Fig. 5.17 The clear price convergence curve using AIS and GA

Table 5.4 Comparison of the solution time using AIS and GA

Solution method	λ^{cl} (CNY/kWh)	Running time (s)
AIS	0.5393	10.011
GA	0.5416	63.088

form a market-based electricity price and power generation allocation plan reflecting MG internal supply and demand and meet load demand.
2. The simulation results in 1.5.3 show that under the proposed model and method, park MG is externally manifested as a friendly and interactive energy unit. The enterprises in MG are sensitive to the electricity price of the main grid and can respond to the purchase and sale electricity, which provides a flexible basis for the price-based demand-side management of large power grids.
3. The simulation results in 1.5.4 show that the AIS-based bidding strategy search method proposed in this chapter can obtain the global optimal solution in the face of complex and unexpected environmental information, and the solution time is only 15.87% of the traditional GA method.

References

1. M. Di Somma, G. Graditi, P. Siano, Optimal bidding strategy for a DER aggregator in the day-ahead market in the presence of demand flexibility. IEEE Trans. Ind. Electron. **66**, 1509–1519 (2019)
2. Z. Xu, P. Yang, C. Zheng, Y. Zhang, J. Peng, Z. Zeng, Analysis on the organization and development of multi-microgrids. Renew. Sust. Energ. Rev. **81**, 2204–2216 (2018)
3. Z. Xu, Y. Zhang, Z. Chen, X. Lin, B. Chen, Bi-level optimal capacity configuration for power to gas facilities considering operation strategy and investment subject benefit. Dianli Xitong Zidonghua/automation of Electric Power Systems **42**, 76–84 (2018)
4. K. Neuhoff, Opening the electricity market to renewable energy: Making better use of the grid. Diw Econ. Bull. **1**, 16–23 (2011)
5. W. Zhu, X. Yuan, Y. Li, Comprehensive Benefit Evaluation Based on Interaction of MicroGrid and Distribution Network, Electr. Energy Manag. Technol. (2014)
6. C. Li, J. Zhang, P. Li, Multi-objective optimization model of micro-grid operation considering cost, pollution discharge and risk. Proc. CSEE **35**, 1051–1058 (2015)
7. M. Rahmani-Andebili, An adaptive approach for PEVs charging management and reconfiguration of electrical distribution system penetrated by renewables, IEEE Trans. Ind. Inf. **14**(5) (2018)
8. L. Jie, L. Yikui, W. Lei, Optimal operation for community-based multi-party microgrid in grid-connected and islanded modes. IEEE Trans. Smart Grid **9**, 756–765 (2018)
9. S. Zhang, L. Shuaikang, Evaluation method of park-level integrated energy system for microgrid. Proc. CSEE **13**, 1983–1996 (2018)
10. D. Gejirifu, Z. Tan, S. Yang, M. Li, B. Yang, W. Kong, N. Zhang, Robust optimization of micro-grid operation considering bidding decisions in electricity market environment, Renew. Energy Res. (2018)
11. K. Shi, N. Liu, J. Zhang, J. Lei, Random matching trading mechanism in microgrid of multi-operators. Power Syst. Technol. (2016)
12. X. Kong, Y. Zeng, N. Lu, J. Yan, Q. Hu, Optimal bidding Management for Agent-based Microgrid Operation. Zhongguo Dianji Gongcheng Xuebao/proceedings of the Chinese Society of Electrical Engineering **37**, 1626–1633 (2017)
13. Mehdi Rahmani-Andebili, Analyzing the Effects of Problem Parameters on the Operation Cost of the Networked Microgrids, 2020 IEEE Kansas Power and Energy Conference (KPEC) (2020)
14. Yongjing, He, Wei, Wang, Xuezhi, and Wu, Multi-Agent Based Fully Distributed Economic Dispatch in Microgrid Using Exact Diffusion Strategy, IEEE Access, 7020–7031 (2019)

15. Z. Zhao, J. Guo, X. Luo, J. Xue, L.L. Lai, Energy transaction for multi-microgrids and internal microgrid based on Blockchain. IEEE Access **76**, 17–29 (2020)
16. M. Jin, W. Feng, P. Liu, C. Marnay, C. Spanos, MOD-DR: Microgrid optimal dispatch with demand response. Appl. Energy **187**, 758–776 (2017)
17. C. Dou, X. Jia, L.I. Heng, Multi-agent-system-based market bidding strategy for distributed generation in microgrid. Power Syst. Technol. (2016)
18. W. Li, Y. Si, L. Ming, J. Zhe, N. Wu, J. Jiang, Microgrid Operational Benefits Distribution Based on the Theory of Cooperative Game, in *2016 China International Conference on Electricity Distribution (CICED)*, (2016)
19. M. Rahmani-Andebili, Chapter 9: Cooperative distributed energy scheduling in Microgrids, in *Electric Distribution Network Management and Control*, (Springer, Singapore, 2018), pp. 235–254
20. T. Wakui, M. Hashiguchi, K. Sawada, R. Yokoyama, Two-stage design optimization based on artificial immune system and mixed-integer linear programming for energy supply networks. Energy **170**, 1228–1248 (2019)
21. A.A. Hafez, A.Y. Hatata, M.M. Aldl, Optimal sizing of hybrid renewable energy system via artificial immune system under frequency stability constraints. J. Renew. Sustain. Energy **11**, 015905 (2019)
22. C. Gao, D. Wu, T. Pan, Z. Ji, Optimal economic dispatch of microgrid based on immune particle swarm optimization algorithm. J. Syst. Simul. **17**, 58–76 (2018)
23. L.W. de Oliveira, E.J. de Oliveira, F.V. Gomes, I.C. Silva, A.L.M. Marcato, P.V.C. Resende, Artificial immune systems applied to the reconfiguration of electrical power distribution networks for energy loss minimization. Int. J. Electr. Power Energy Syst. **56**, 64–74 (2014)
24. B. Ramachandran, S.K. Srivastava, C.S. Edrington, D.A. Cartes, An intelligent auction scheme for Smart grid market using a hybrid immune algorithm. IEEE Trans. Ind. Electron. **58**, 4603–4612 (2011)
25. L. Tao, C. Schwaegerl, S. Narayanan, J.H. Zhang, From Laboratory Microgrid to Real Markets — Challenges and Opportunities, in *IEEE International Conference on Power Electronics & Ecce Asia* (2011)
26. S. Parhizi, A. Khodaei, M. Shahidehpour, Market-based versus Price-based microgrid optimal scheduling. IEEE Trans. Smart Grid **9**, 615–623 (2018)
27. J.K. Liu, E.R. Lian-Jie, Overview of application of multiagent technology, Control & Decision (2001)
28. M. Rahmani-Andebili, Chapter 4: Robust operation of a reconfigurable electrical distribution system by optimal charging management of plug-in electric vehicles considering the technical, social, and geographical aspects, in *Planning and Operation of Plug-In Electric Vehicles: Technical, Geographical, and Social Aspects*, (Springer, Cham, 2019)
29. M. Rahmani-Andebili, Dynamic and adaptive reconfiguration of electrical distribution system including renewables applying stochastic model predictive control. IET Gener. Trans. Distrib. **11**(16), 3912–3921 (2017)
30. Tan and Ying, *Artificial Immune System* (Springer, Berlin, 2016)
31. X. Fang, J. Li, L. Li, Restudying the Artificial Immune Model for Network Intrusion Detection, in *2009 International Workshop on Intelligent Systems and Applications* Wuhan,China (2009)
32. X. Kong, D. Liu, J. Xiao, C. Wang, A multi-agent optimal bidding strategy in microgrids based on artificial immune system. Energy **189**, 116154 (2019)

Chapter 6
Microgrid Formation Strategy Including Multiple Energy and Capacity Resources for Resilience Improvement

Hasan Mehrjerdi, Sajad Mahdavi, and Reza Hemmati

Abstract Nowadays, the resilience enhancement is one of the most important concerns in electric power networks. The division of the main microgrid into several sub-microgrids, i.e., microgrid formation (MF), is a resilient strategy for distribution systems against natural disasters and cyber-physical attacks. Such effective solution not only increases the resilience and load restoration but also reduces the costs. The extensive penetration of renewable resources in microgrids increases the issues about safe operation under faults. This chapter presents a resilient microgrid formation in the presence of solar, wind, and diesel Distributed generation (DG) for load restoration maximization. In order to carry out the microgrid formation, several candidate breakers and tie-line switches are considered, and their optimal on-off conditions are determined. Both the active and reactive powers are included in the model. The model is expressed as mixed-integer linear programming (MILP) and is simulated under three various cases including case 1, without formation strategy; case 2, formation strategy with line breaker switch; and case 3, formation strategy with both line and tie breaker switches. The numerical results are carried out based on IEEE 33-bus and 69-bus standard distribution networks. The results emphasize on the effectiveness of the developed formation strategy with both the breaker and tie line switches for load restoration and resilience enhancement.

Keywords Fictitious network · Load restoration · Microgrid formation · Piecewise linearization · Resilience · Renewable resources

H. Mehrjerdi (✉)
Electrical Engineering Department, Qatar University, Doha, Qatar
e-mail: Hasan.mehrjerdi@qu.edu.qa

S. Mahdavi · R. Hemmati
Department of Electrical Engineering, Kermanshah University of Technology, Kermanshah, Iran
e-mail: r.hemmati@kut.ac.ir

M. Rahmani-Andebili (ed.), *Design, Control, and Operation of Microgrids in Smart Grids*, Power Systems, https://doi.org/10.1007/978-3-030-64631-8_6

Nomenclature

Indices and sets	
i,j	Index of network buses
l	Index of piecewise lines for linearization
ψ_b	Set of network buses
ψ_{dg}	Set of DGs
Parameters	
g_{ij}, b_{ij}	Conductance and susceptance between buses i and j (p.u.)
PD, QD	Active and reactive power demand (kW, kVar)
P^s, P^w	Active power of solar and wind units (kW)
L	Number of all piecewise lines for linearization
M	Big digit in big-m inequalities
N_b, N_{sw}, N_{dg}	Number of network buses, tie switches, and DGs
S^{max}, S^{dg}	Maximum apparent power of line and DG (kVA)
θ^{max}	Maximum angle difference between buses of network (rad)
α_i	Power factor of load
ω_i	Weight of load
Δv_{max}	Maximum difference of bus voltage from 1 p.u.
Binary variables	
ξ	Control variable of the breaker switch
δ, λ	Auxiliary variables for linearization of θ^2
Continuous variables	
P_{ij}, Q_{ij}	Active and reactive powers between buses i and j (kW, kVar)
$\widetilde{P_{ij}}$, $\widetilde{P_{dg}}$	Active power between buses i and j and active power of DG (kW)
P_{ij}^{loss}, Q_{ij}^{loss}	Active and reactive losses between buses i and j (kW, kVar)
pd_i, qd_i	Curtailed active and reactive loads at bus i (kW, kVar)
P_i^{dg}, Q_i^{dg}	Active and reactive powers of DG at bus i (kW, kVar)
v_i, θ_i	Voltage magnitude and angle of bus i (p.u., rad)
Δv_i	Voltage magnitude deviation from 1 (p.u.)
θ_{ij}	Voltage angle difference between buses i and j (rad)
$\Delta\theta_{ij}$, θ_{ij}^+, θ_{ij}^-	Auxiliary variables for linearization of θ^2

6.1 Introduction

6.1.1 Motivations of This Work

In recent years, climate change and natural disasters (e.g., flood, earthquake, storm, and ice storm) have intensified in many countries [1]. The extreme and costly outages due to natural disasters and attacks [2] have attracted more attention on the resilience concept. The statistics and reports show that the outages have increased more than 200% over the last years and more than 80% of these faults happen in the distribution networks. The distribution network outages make significant effects on the quality of electricity and customer satisfaction [3]. For example, on October 29, 2012, more than 7.5 million people were left without electricity during Hurricane Sandy. On December 23, 2013, the cyber-attack to several energy companies caused a 6-h power outage for 230,000 people in Ukraine [4]. The violent storm in South Australia on September 28, 2016, disconnected power of 850 thousand customers [5]. Hurricane Irma on September 10, 2017, switched off power to 6.7 million customers in Florida [6]. In view of the above contents, the natural disasters and attacks are the serious threats for electrical infrastructures [7]. Therefore, due to the extreme vulnerability of distribution networks [8], it seems very necessary and vital to provide a proper framework to reduce the effects of failures, rapid load restoration, and resilience improvement under critical situations.

6.1.2 Literature Survey and Contributions

The total or partial loss of generation due to high load demand, disconnecting network facilities, cyber-attacks, and natural disasters are the important factors for network instability, system collapse, and widespread blackouts [9]. Each event not only affects the people but also causes billion-dollar economic damages to the system [10]. Therefore, researchers have focused on resilience enhancement following disasters to achieve resilient power networks [11, 12]. The efficient and rapid restoration is one of the key stages for resilience enhancement in distribution networks [11]. When a fault happens in the distribution substation, the substation cannot be supplied by the main grid, and it may be separated from the other areas. In this state, the traditional methods of network restoration cannot guarantee the uninterrupted supply of loads [13]. One of the efficient methods to remove such issues is changing the structure of the network and splitting the network into several sub-microgrids by line and tie switches (i.e., microgrid formation). Therefore, the microgrid formation or deliberate islanding of microgrid is a promising strategy for load restoration subject to satisfaction of all technical constraints in every sub-microgrid [14].

The microgrid formation with the aim of maximizing load restoration has been studied by many literatures based on various methods like expert system [15, 16], fuzzy logic [17, 18], multi-agent systems [19, 20], heuristic search [21], and optimization [22]. The microgrid formation considering power losses and voltage constraints in order to maximize load restoration is an effective model [23]. Such problems are often modelled as mixed-integer nonlinear programs (MINLP) where the status of line and tie switches for sectionalizing the network are optimized. The distributed generators (DGs) can make distribution networks more flexible and secure [24]. The DGs can be used to enhance load restoration and provide an alternative way to improve resilience in grids [13] by dividing the network into smaller clusters or sub-grids [25]. The reconfiguration or sectionalizing the microgrid into several sub-microgrids which are energized by DGs can extremely increase load restoration. The model given by [24] presents a microgrid formation based on master-slave diesel DG in which every sub-microgrid is installed with one master diesel. The aforementioned model shows that the microgrid reconfiguration by both the line and tie switches leads to more load restoration. The microgrid formation by MILP is very computational for large networks [26]. As a result, the heuristic methods are often used for large-scale systems [27]. A framework for both the radial and meshed systems in order to maximize load restoration is presented by [28]. The authors in [29] present a master-slave DG technique for resilience enhancement by microgrid formation and optimal diesel DG integration. The authors in [30] apply the minimum spanning tree search algorithm for microgrid formation in order to critical load restoration based on graph theory. According to the aforementioned model, the main aim of load restoration in the distribution network is to maximize the customer survivability by load shedding minimization and guaranteeing the network radiality. A two-stage algorithm based on nested microgrid formation for minimizing the unused capacity of diesel and self-healing restoration is proposed by [10]. The model demonstrates that the power sharing between the microgrids is one of the solutions to increase the unused capacity of diesel DGs. The authors in [6] identify the optimal microgrid formation and resilience maximization when the system is damaged. The graph theory and spanning tree are also proper methods for microgrid formation [31]. The microgrid planning considering critical load restoration and uncertain formation can also be addressed for cost minimization [32]. A multi-microgrid strategy based on system of systems is presented by [33]. This work studies the resilience and energy management in multi-microgrid system. In the proposed model, the microgrid is formed by four sub-microgrids. A zonal formation strategy is presented by [34] in order to network smallness and computational time reduction. The battery energy storages and electric vehicles may also be integrated for reducing load shedding [35, 36].

This chapter addresses the microgrid formation strategy subject to necessary conditions for network radiality. The main purpose of this chapter is to study the microgrid formation in order to load restoration and resilience maximization at fault time. Therefore, the time horizon and parametric uncertainty are not included in the proposed model. The developed model is a deterministic planning. One of the best methods of uncertainty modelling is stochastic model predictive control (stochastic MPC) where details of this strategy can be found in [37–41]. The problem is simulated as MILP in order to load restoration maximization considering three various cases. The open/close states of line and tie switches are optimized, while all constraints such as radiality and connectivity are satisfied. The main contributions of the developed model can be summarized as follows:

- The microgrid formation is presented considering only one diesel DG in each island section to guarantee the island self-adequacy.
- The linear power flow is presented by employment of piecewise linear approximation including active/reactive losses and voltage phasor.
- Both the line and tie switches are considered in order to have more diversity in system topology.
- The load in one feeder can be transferred to another feeder by tie breaker switches in order to have more load restoration.

6.2 Microgrid Formation Problem

As mentioned in the previous section, sectionalizing the microgrid into several smaller microgrids is one of the main solutions to increase the resilience through increasing the load restoration level. In order to guarantee the self-adequacy in each sub-grid, there is at least one DG in each island section, and the voltage magnitude in the DG location is fixed on 1 p.u. In this strategy, each DG has only the task of supplying the loads close to itself instead of suppling the loads in farther areas. The main constraints of microgrid formation can be summarized as follows:

- Radiality and connectivity of island under any condition
- Open/close state of breaker switches
- Establishment of power flow equations in each island
- Other constraints such as line capacity, DG capacity, and bus voltage range

Generally, in microgrid formation, there are three kinds of binary variables:

- Open/close state of each line by both line and tie-line breaker switches
- Open/close state of load breaker switch
- Open/close state of DG breaker switch

With respect to the number of binary variables in the model, the problem complexity and computational time are high. In the practical networks, all lines are not allowed to be opened because of blackout or stability constraints. As a result, the breaker switches are only installed on some candidate lines. In addition, the loads can be considered as dispatchable and be curtailed totally or partially. Such model uses the continuous variables instead of binary ones. Based on such models, the number of binary variables in the model can be reduced [23]. The connectivity constraints for each island are very essential for power systems because such constraints reduce the number of islands and increase the system reliability. Without these constraints, it is not guaranteed to have subgraphs since the subgraphs may be disconnected [42] as shown in Fig. 6.1.

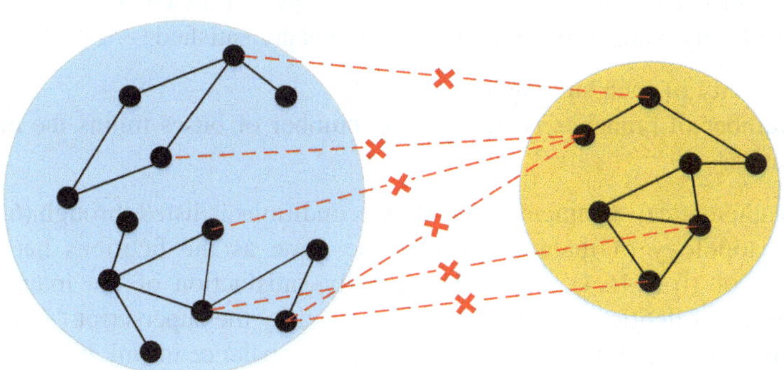

Fig. 6.1 Non-connectivity of divided subgraphs

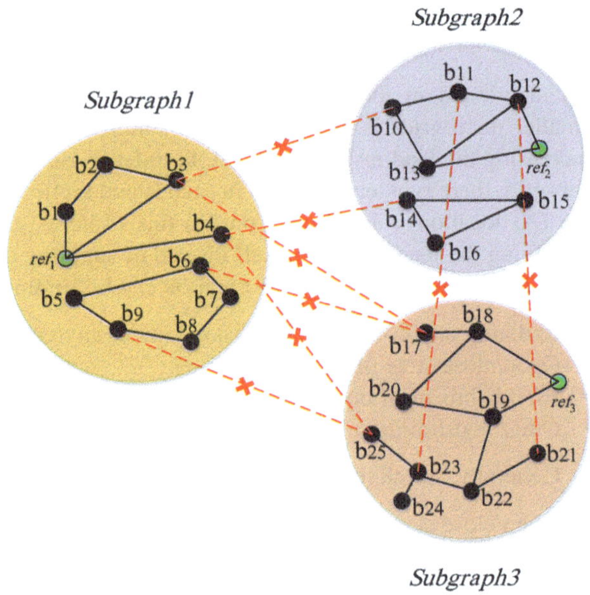

Fig. 6.2 Connectivity constraints of the fictitious network

The authors in [43] have shown that in order to guarantee such constraints, many binary variables are required. Therefore, in order to reduce the binary variables and avoid complexity, a single-commodity power flow is proposed by [42]. First, the fictitious network is designed where each subgraph is powered by only one source, known as a "source," and all other buses have unit load demand (1 p.u.), known as a "sink." Each "source" can be a reference bus with voltage amplitude equal to 1 p.u. The satisfaction of power balance on each bus needs at least one path between the bus and the "source." The subgraph connectivity is guaranteed if the power balance is achieved in the subgraph. As seen in Fig. 6.2, the power in subgraphs 1 and 2 cannot be balanced because there is no line between b_5 and b_9 with ref_1 and between b_{14} and b_{16} with ref_2 but the power in subgraph 3 is balanced.

A radial distribution system must remain radial after formation; therefore, the sufficient and necessary conditions of radiality are expressed by [44]. A graph is radial if and only if the two following conditions are satisfied:

- Connectivity of each subgraph is satisfied.
- The number of branches is equal to the number of buses minus the number of subgraphs.

The mathematical statement of the above conditions is listed through (6.1)–(6.4). Since the topology of main network is the same as the fictitious network, the satisfaction of (6.1)–(6.4) means the radiality satisfaction of the main network. The variables of fictitious power flow are marked by the superscript "~."

Equations (6.1) and (6.2) show the active power balance in sink buses and source buses, respectively. Open and close state of each line is expressed by (6.3) where $\xi = 1$ means line is closed; otherwise, line is open and the active flow is equal to zero. In each radial network, the number of closed lines is equal to the number of buses minus the number of tie switches and DGs; otherwise, the mesh is created in network. This issue is confirmed by (6.4).

$$\sum_{i \neq j} \widetilde{P}_{ij} = -1 \qquad \forall\ i \in \psi_b \backslash \psi_{dg} \tag{6.1}$$

$$\sum_{i \neq j} \widetilde{P}_{ij} = -\widetilde{P}_{DG} \qquad \forall\ i \in \psi_{dg} \tag{6.2}$$

$$-\zeta_{ij} M \leq \widetilde{P}_{ij} \leq \zeta_{ij} M \qquad \forall i, j \in \psi_b \tag{6.3}$$

$$\sum_{i \neq j} \xi_{ij} = N_b - N_{sw} - N_{dg} \quad \forall i, j \in \psi_b \tag{6.4}$$

6.3 Mathematical Formulation

6.3.1 Power Flow Equations

The power flow direction between buses i and j is shown in Fig. 6.3. The basic equation of apparent power related to this system is expressed by (6.5), where (6.6)–(6.7) are achieved after separating the real and imaginary parts. These relationships are extremely nonlinear. In order to linearize the equations, the linearization assumptions shown by (6.8) are considered. After elimination of very small parts, the active and reactive powers and losses between buses i and j are calculated through (6.9) to (6.12).

$$P_{ij} - jQ_{ij} = v_i^* (v_i - v_j) / (g_{ij} + jb_{ij}) \tag{6.5}$$

$$P_{ij} = v_i^2 g_{ij} - v_i v_j (g_{ij} \cos \theta_{ij} + b_{ij} \sin \theta_{ij}) \tag{6.6}$$

$$Q_{ij} = -v_i^2 b_{ij} - v_i v_j (g_{ij} \sin \theta_{ij} - b_{ij} \cos \theta_{ij}) \tag{6.7}$$

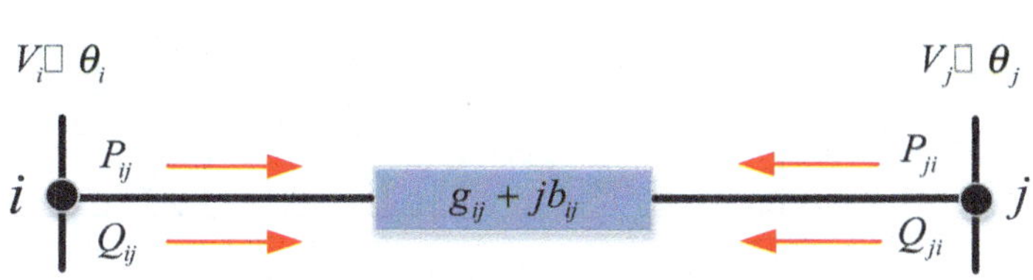

Fig. 6.3 Power flow direction between buses i and j

$$\text{Linearization asumptions}\begin{cases} v_i = 1 + \Delta v_i \\ \sin\theta_{ij} = \theta_{ij} \\ \cos\theta_{ij} = 1 - \theta_{ij}^2/2 \end{cases} \tag{6.8}$$

$$P_{ij} = g_{ij}\left(\Delta v_i - \Delta v_j\right) - b_{ij}\theta_{ij} + g_{ij}\theta_{ij}^2/2 \tag{6.9}$$

$$Q_{ij} = -b_{ij}\left(\Delta v_i - \Delta v_j\right) - g_{ij}\theta_{ij} - b_{ij}\theta_{ij}^2/2 \tag{6.10}$$

$$P_{ij}^{loss} = g_{ij}\theta_{ij}^2 \tag{6.11}$$

$$Q_{ij}^{loss} = -b_{ij}\theta_{ij}^2 \tag{6.12}$$

6.3.2 Second-Order Curve Linearization

As seen in Eqs. (6.9) to (6.12), the term "θ^2" is a nonlinear term, and this second-order curve is linearized by applying the piecewise linear approximation method. Figure 6.4 shows the overview of this method that is expressed by (6.13)–(6.20). The details of this strategy can be found in [45].

$$\theta_{ij}^2 = \sum_{l \in L}(2l-1)\Delta\theta_{ij}^l.\theta^{\max}/L \tag{6.13}$$

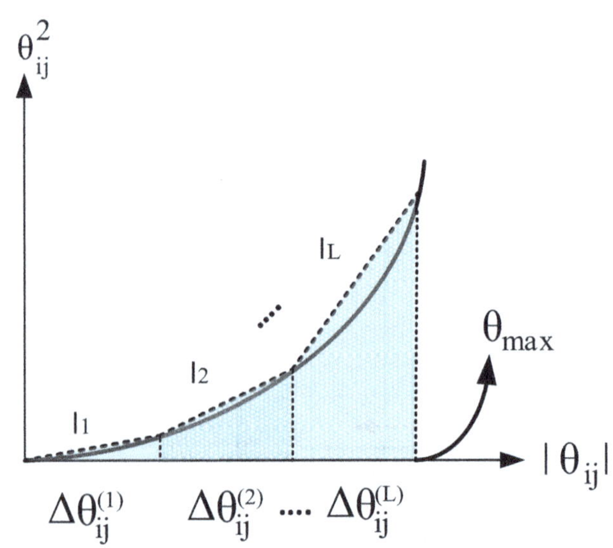

Fig. 6.4 Piecewise linear approximation of the second-order curve

$$\theta_{ij} = \theta_{ij}^{+} - \theta_{ij}^{-} \tag{6.14}$$

$$\sum_{l \in L} \Delta\theta_{ij}^{l} = \theta_{ij}^{+} + \theta_{ij}^{-} \tag{6.15}$$

$$\theta_{ij}^{+} \leq \delta_{ij}\theta^{\max} \tag{6.16}$$

$$\theta_{ij}^{-} \leq (1 - \delta_{ij})\theta^{\max} \tag{6.17}$$

$$\lambda_{ij}^{l}.\theta^{\max}/L_{\max} \leq \Delta\theta_{ij}^{l} \leq \theta^{\max}/L \qquad \forall l = 1 \tag{6.18}$$

$$\lambda_{ij}^{l}.\theta^{\max}/L_{\max} \leq \Delta\theta_{ij}^{l} \leq \lambda_{ij}^{1-l}\theta^{\max}/L \qquad \forall\ 1 < l < L \tag{6.19}$$

$$0 \leq \Delta\theta_{ij}^{l} \leq \lambda_{ij}^{l-1}.\theta^{\max}/L \qquad \forall l = L \tag{6.20}$$

6.4 Reformed Network Equations

6.4.1 Line Flows and Losses

After calculation of linearized equations of the problem, the line flow and line losses of the reformed network are achieved by Big-M inequalities listed in (6.21)–(6.28), where $\xi_{ij} = 1$ means the breaker switch between buses i and j is closed; otherwise, the breaker switch is open.

$$-\xi_{ij}M \leq P_{ij} \leq \xi_{ij}M \tag{6.21}$$

$$-(1 - \xi_{ij})M \leq P_{ij} - g_{ij}(\Delta v_i - \Delta v_j) + b_{ij}\theta_{ij} - g_{ij}\theta_{ij}^2/2 \leq (1 - \xi_{ij})M \tag{6.22}$$

$$-\xi_{ij}M \leq Q_{ij} \leq \xi_{ij}M \tag{6.23}$$

$$-(1 - \xi_{ij})M \leq Q_{ij} + b_{ij}(\Delta v_i - \Delta v_j) + g_{ij}\theta_{ij} + b_{ij}\theta_{ij}^2/2 \leq (1 - \xi_{ij})M \tag{6.24}$$

$$-\xi_{ij}M \leq P_{ij}^{\text{loss}} \leq \xi_{ij}M \tag{6.25}$$

$$-(1 - \xi_{ij})M \leq P_{ij}^{\text{loss}} - g_{ij}\theta_{ij}^2 \leq (1 - \xi_{ij})M \tag{6.26}$$

$$-\xi_{ij}M \leq Q_{ij}^{\text{loss}} \leq \xi_{ij}M \tag{6.27}$$

$$-(1 - \xi_{ij})M \leq Q_{ij}^{loss} + b_{ij}\theta_{ij}^2 \leq (1 - \xi_{ij})M \tag{6.28}$$

6.4.2 Circle Linearization

The line capacity constraint is presented by nonlinear inequality (6.29), and in order to linearize this model, the approximation of a circle by enclosed hexagon is adopted as shown in Fig. 6.5. In this method, L is the number of hexagon sides, and inequalities shown in (6.30) are used instead of those in (6.29). The details of the method are addressed by [46, 47].

$$P_{ij}^2 + Q_{ij}^2 \leq S_{ij}^{\max\,2} \tag{6.29}$$

$$P_{ij}.\cos\frac{(2l-1)}{L}\pi + Q_{ij}.\sin\frac{(2l-1)}{L}\pi - S_{ij}^{\max}\cos\frac{\pi}{L} \leq 0 \quad \forall l \in \{1,2,\ldots L\} \tag{6.30}$$

The DG capacity constraint is calculated by (6.31), and similarly, the inequalities listed in (6.32) are applied for linearizing (6.31).

$$P_i^{dg^2} + Q_i^{dg^2} \leq S_i^{dg^2} \qquad \forall i \in \psi_{dg} \tag{6.31}$$

$$P_i^{dg}.\cos\frac{(2l-1)}{L}\pi + Q_i^{dg}.\sin\frac{(2l-1)}{L}\pi - S_i^{dg}\cos\frac{\pi}{L} \leq 0 \quad \forall l \in \{1,2,\ldots L\} \tag{6.32}$$

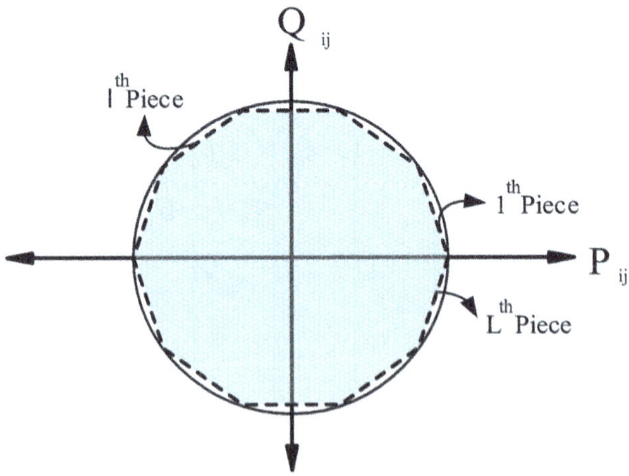

Fig. 6.5 Approximation of a circle by an enclosed hexagon

6.4.3 Power Balance Constraints

The active and reactive power balance equations are calculated using (6.33) and (6.34). As mentioned in the previous sections, in order to reduce the number of binary variables, the load is modelled as dispatchable and can be partially or totally curtailed. The dispatchability of load is presented in (6.35) and (6.36) where the curtailed reactive part is proportional to the curtailed active part and power factor angle. In the buses integrated with DG, the voltage magnitude is set to 1 p.u. as presented by (6.37) because these buses are the "source" buses in each island section. The other buses are the "sink" buses, and the voltage magnitude may be variable inside the permitted range as shown by (6.38).

$$\sum_{j \neq i} P_{ij} + pd_i - P_i^w - P_i^s - P_i^{dg} = 0 \qquad \forall i, j \in \psi_b \tag{6.33}$$

$$\sum_{j \neq i} Q_{ij} + qd_i - Q_i^{dg} = 0 \qquad \forall i, j \in \psi_b \tag{6.34}$$

$$0 \leq pd_i \leq PD_i \tag{6.35}$$

$$qd_i = pd_i \tan \alpha_i \tag{6.36}$$

$$\Delta v_i = 0 \qquad \forall i \in \psi_{dg} \tag{6.37}$$

$$-\Delta v_{\max} \leq \Delta v_i \leq \Delta v_{\max} \qquad \forall i \in \psi_b \backslash \psi_{dg} \tag{6.38}$$

6.4.4 Objective Function

The objective function of the problem (i.e., load restoration) subject to the required constraints is presented by (6.39) that must be maximized, where ω_i is the coefficient showing priority or importance of each load. Some loads such as hospitals and fire stations are persuasive, and ω_i is higher for these loads.

$$\begin{cases} Max: \sum \omega_i pd_i & \forall i \in \psi_b \\ \text{s.t.} \quad (1)\text{-}(4),\ (13)\text{-}(28),\ (30),\ (32)\text{-}(38) \end{cases} \tag{6.39}$$

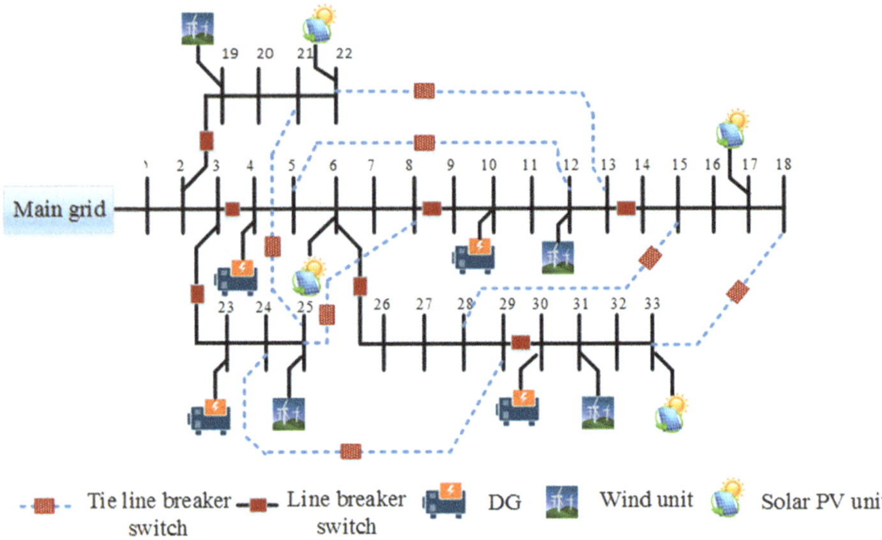

Fig. 6.6 Modified IEEE 33-bus distribution network

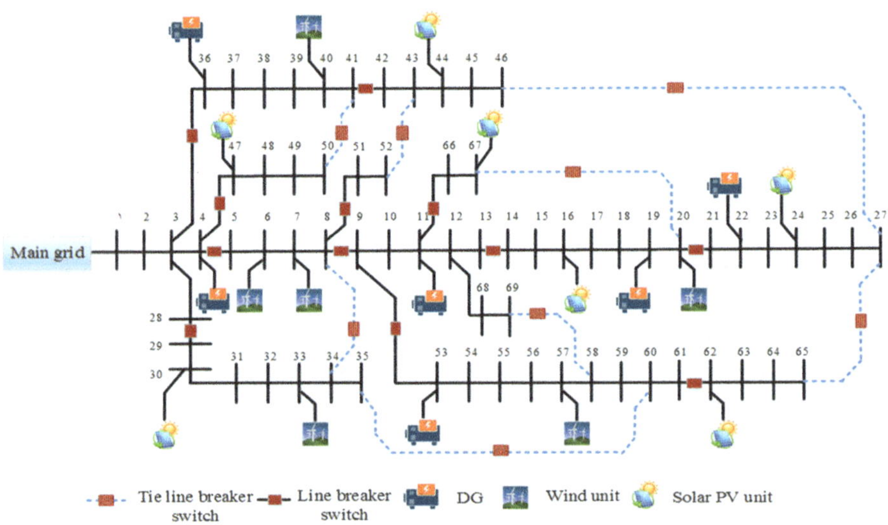

Fig. 6.7 Modified IEEE 69-bus distribution network

6.5 Case Studies

The structure of modified IEEE 33- and 69-bus distribution networks is shown in Figs. 6.6 and 6.7, respectively. The line and load data of networks are listed in Table 6.1 [48, 49]. The total load of the first test system is 3715 kW and 2300 kVar,

Table 6.1 Line and load data for IEEE 33- and 69-bus networks

33-Bus network						69-Bus network					
Line location		Line parameters		Bus loads at receiving bus		Line location		Line parameters		Bus loads at receiving bus	
Sending bus	Receiving bus	R (Ω)	X (Ω)	P (kW)	Q (kVar)	Sending bus	Receiving bus	R (Ω)	X (Ω)	P (kW)	Q (kVar)
1	2	0.0922	0.047	100	60	1	2	0.0005	0.0012	0.0	0.0
						2	3	0.0005	0.0012	0.0	0.0
2	3	0.493	0.2511	90	40	3	4	0.0015	0.0036	0.0	0.0
						4	5	0.0251	0.0294	0.0	0.0
3	4	0.366	0.1864	120	80	5	6	0.3660	0.1864	2.60	2.20
						6	7	0.3811	0.1941	40.40	30.00
4	5	0.3811	0.1941	60	30	7	8	0.0922	0.0470	75.00	54.00
						8	9	0.0493	0.0251	30.00	22.00
5	6	0.819	0.707	60	20	9	10	0.8190	0.2707	28.00	19.00
						10	11	0.1872	0.0619	145.00	104.00
6	7	0.1872	0.6188	200	100	11	12	0.7114	0.2351	145.00	104.00
						12	13	1.0300	0.3400	8.00	5.00
7	8	0.7114	0.2351	200	100	13	14	1.0440	0.3450	8.00	5.50
						14	15	1.0580	0.3496	0.0	0.0
8	9	1.03	0.74	60	20	15	16	0.1966	0.0650	45.50	30.00
						16	17	0.3744	0.1238	60.00	35.00
9	10	1.044	0.74	60	20	17	18	0.0047	0.0016	60.00	35.00
						18	19	0.3276	0.1083	0.0	0.0
10	11	0.1966	0.065	45	30	19	20	0.2106	0.0690	1.00	0.60
						20	21	0.3416	0.1129	114.00	81.00
11	12	0.3744	0.1238	60	35	21	22	0.0140	0.0046	5.00	3.50
						22	23	0.1591	0.0526	0.0	0.0

(continued)

Table 6.1 (continued)

33-Bus network						69-Bus network					
Line location		Line parameters		Bus loads at receiving bus		Line location		Line parameters		Bus loads at receiving bus	
Sending bus	Receiving bus	R (Ω)	X (Ω)	P (kW)	Q (kVar)	Sending bus	Receiving bus	R (Ω)	X (Ω)	P (kW)	Q (kVar)
12	13	1.468	1.155	60	35	23	24	0.3463	0.1145	28.00	20.0
						24	25	0.7488	0.2475	0.0	0.0
13	14	0.5416	0.7129	120	80	25	26	0.3089	0.1021	14.0	10.0
						26	27	0.1732	0.0572	14.0	10.0
14	15	0.591	0.526	60	10	3	28	0.0044	0.0108	26.0	18.6
						28	29	0.0640	0.1565	26.0	18.6
15	16	0.7463	0.545	60	20	29	30	0.3978	0.1315	0.0	0.0
						30	31	0.0702	0.0232	0.0	0.0
16	17	1.289	1.721	60	20	31	32	0.3510	0.1160	0.0	0.0
						32	33	0.8390	0.2816	14.0	10.0
17	18	0.732	0.574	90	40	33	34	1.7080	0.5646	9.50	14.00
						34	35	1.4740	0.4873	6.00	4.00
2	19	0.164	0.1565	90	40	3	36	0.0044	0.0108	26.0	18.55
						36	37	0.0640	0.1565	26.0	18.55
19	20	1.5042	1.3554	90	40	37	38	0.1053	0.1230	0.0	0.0
						38	39	0.0304	0.0355	24.0	17.00
20	21	0.4095	0.4784	90	40	39	40	0.0018	0.0021	24.0	17.00
						40	41	0.7283	0.8509	1.20	1.0
21	22	0.7089	0.9373	90	40	41	42	0.3100	0.3623	0.0	0.0
						42	43	0.0410	0.0478	6.0	4.30
3	3	0.4512	0.3083	90	50	43	44	0.0092	0.0116	0.0	0.0
						44	45	0.1089	0.1373	39.22	26.30
23	24	0.898	0.7091	420	200	45	46	0.0009	0.0012	39.22	26.30
						4	47	0.0034	0.0084	0.00	0.0

24	25	0.896	0.7011	420	200	47	48	0.0851	0.2083	79.00	56.40
						48	49	0.2898	0.7091	384.70	274.50
6	26	0.203	0.1034	60	25	49	50	0.0822	0.2011	384.70	274.50
						8	51	0.0928	0.0473	40.50	28.30
26	27	0.2842	0.1447	60	25	51	52	0.3319	0.1114	3.60	2.70
						52	53	0.1740	0.0886	4.35	3.50
27	28	1.059	0.9337	60	20	53	54	0.2030	0.1034	26.40	19.00
						54	55	0.2842	0.1447	24.00	17.20
28	29	0.8042	0.7006	120	70	55	56	0.2813	0.1433	0.0	0.0
						56	57	1.5900	0.5337	0.0	0.0
29	30	0.5075	0.2585	200	600	57	58	0.7837	0.2630	0.0	0.0
						58	59	0.3042	0.1006	100.0	72.0
30	31	0.9744	0.963	150	70	59	60	0.3861	0.1172	0.0	0.0
						60	61	0.5075	0.2585	1244.0	888.00
31	32	0.3105	0.3619	210	100	61	62	0.0974	0.0496	32.0	23.00
						62	63	0.1450	0.0738	0.0	0.0
32	33	0.341	0.5302	60	40	63	64	0.7105	0.3619	227.0	162.00
						64	65	1.0410	0.5302	59.0	42.0
						11	66	0.2012	0.0611	18.0	13.0
						66	67	0.0047	0.0014	18.0	13.0
						12	68	0.7394	0.2444	28.0	20.0
						68	69	0.0047	0.0016	28.0	20.0

Table 6.2 .Significant level of loads

33-Bus network		69-Bus network	
Bus no.	Weight	Bus no.	Weight
1, 2, 3	1	1, 2, 3, 4, 5, 6, 7, 8, 9, 10, 11	1
4, 5, 6	2.6	12, 13, 14, 15, 16, 17, 18, 19	5
7, 8, 9, 10	3	20, 21, 22, 23, 24, 25, 26, 27	2.5
11, 12, 13	5	28, 29, 30	1
14, 15, 16	1.5	31, 32, 33, 34, 35, 36, 37, 38, 39	3.5
17, 18	1.8	40, 41, 42, 43, 44, 45, 46	2
19, 20, 21, 22	1	47, 48, 49, 50, 51, 52	5
23, 24, 25	5	53, 54, 55, 56, 57	4
26, 27, 28	3	58, 59, 60	5
29, 30	4.2	61, 62, 63, 64, 65	4.5
31, 32, 33	2	66, 67, 68, 69	2

while the total load of the second grid is 3802 kW and 2695 kVar. In both networks, the wind and solar PV are installed on some of the buses with a 300-kW nominal power. In the 33-bus distribution network, the number of line and tie breaker switches is 7, while in the other network, the number of these switches is 12 and 8, respectively. The status of line and tie breaker switches in the normal condition is called "normally"; for instance, some switches are "normally closed" and the rest are "normally open." The line breakers are normally closed, and the tie-line breakers are normally open.

The nominal capacity of DGs is 400 KVA. The DGs not only can generate active power but also can generate or absorb reactive power according to DG power factor. As mentioned in previous sections, after microgrid formation, the number of islands is equal to the number of DGs which means four and six islands in the first and second test systems, respectively. The voltage magnitude on the DG buses is set to 1 per unit. The significant level or weight of loads is different in each distribution network. In the load restoration strategy, the loads with higher weight have priority to be dealt with. Table 6.2 shows the load weight in both 33- and 69-bus networks. As seen in the table, this level is variable from 1 (low importance) to 5 (high importance).

The CPLEX solver [50] within the GAMS [51] environment is used to implement and solve the given formulation.

6.6 Simulation Results and Discussions

The model is simulated under three cases. The first one is load restoration maximization without formation, the second one is formation strategy without tie line breaker switch, and the third one is formation strategy with tie line breaker switch. In all the cases, it is assumed that a single fault occurs in the main transformer that causes microgrid isolation from the main grid. In order to satisfy the power flow formulation under case 1, the voltage magnitude on bus 4 (in 33-bus network) and bus 10 (in 69-bus network) is set to 1 per unit. In the other cases, the voltage magnitude of DG buses is set to 1 per unit.

6.6.1 Load Restoration Under Different Cases

The structure of reformed 33- and 69-bus networks under cases 2 and 3 is shown in Figs. 6.8, 6.9, 6.10, and 6.11. As seen in the figures, the radiality and connectivity constraints are satisfied, and the number of the created islands is equal to the number of DGs. The load restoration (i.e., objective function) under three cases is listed in Table 6.3. In case 1 (without formation), the load restoration is 52% and 22% for

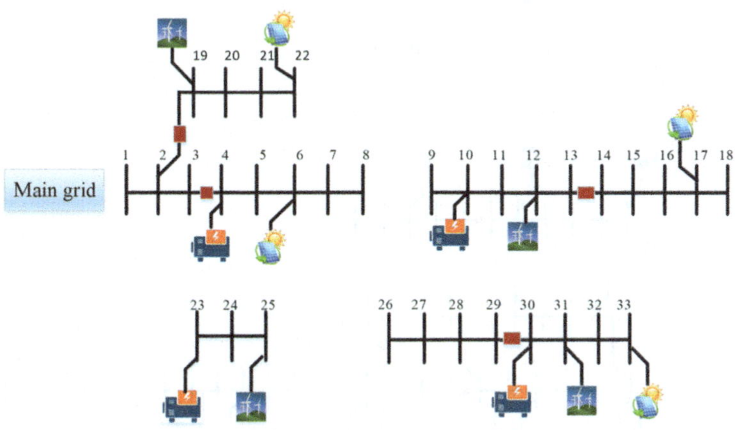

Fig. 6.8 Structure of reformed 33-bus network in case 2

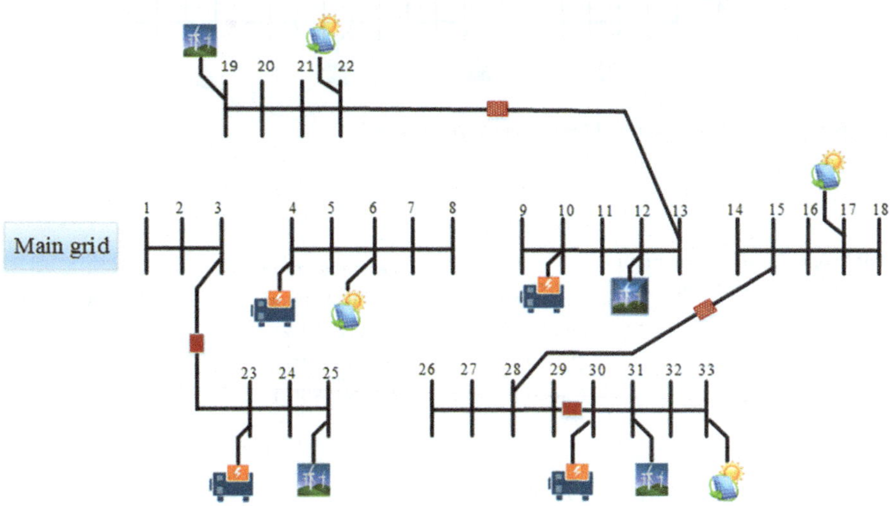

Fig. 6.9 Structure of reformed 33-bus network in case 3

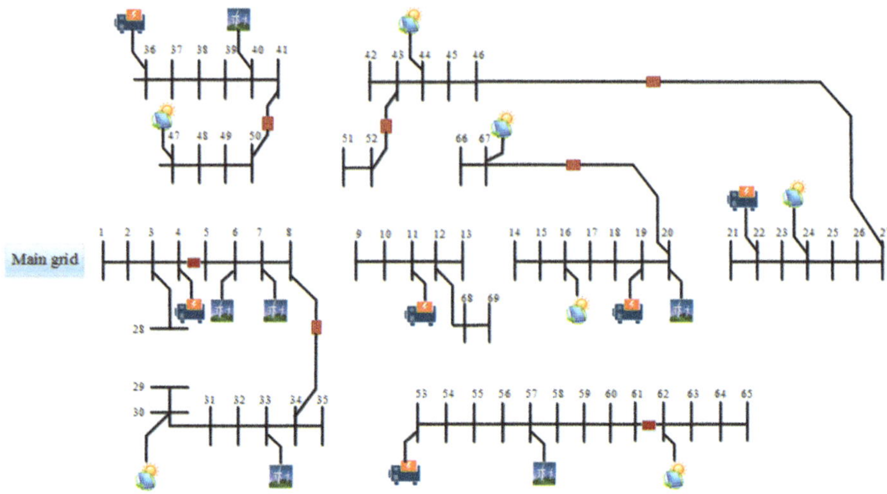

Fig. 6.11 Structure of reformed 69-bus network in case 3

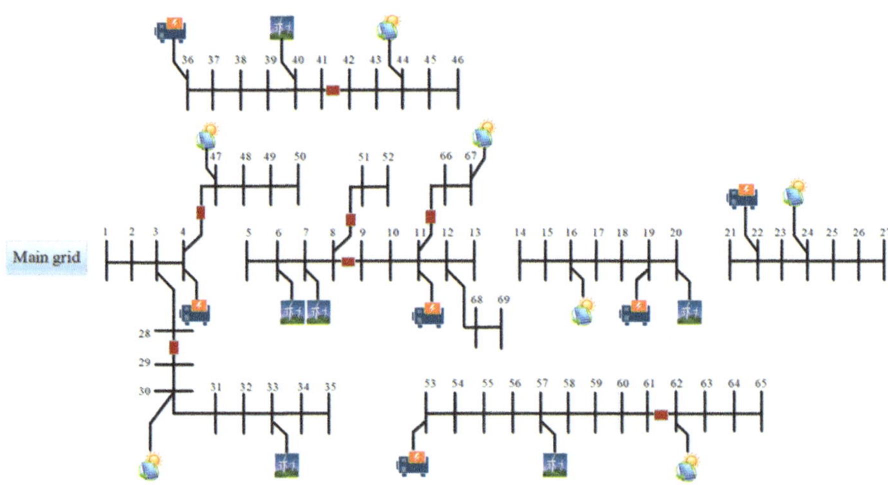

Fig. 6.10 Structure of reformed 69-bus network in case 2

Table 6.3 Load restoration in networks under different cases

Case no.	Load restoration (kW)	
	33-bus network	69-bus network
Case 1	5849 (52%)	3010 (22%)
Case 2	8304 (74%)	8014 (58%)
Case 3	10,260 (89%)	13,121 (95%)

both networks, while this index is increased in the other cases. According to the results, in the 33-bus network, the loads are restored by 74% and 89% for cases 2 and 3, respectively. In addition, in the 69-bus network, the load restoration is 58% and 95% for cases 2 and 3, respectively.

In the formation strategy, each DG is responsible to supply its local load. Therefore, the load restoration is improved in cases 2 and 3 compared to case 1. In addition, the tie lines enable the system to transfer the load from one feeder to another feeder, and this point increases the load restoration in case 3 compared to case 2.

The active loading of 33 and 69 bus networks under different cases is shown in Figs. 6.12 and 6.13. In case 1, the low significant loads are curtailed in the 33-bus network such as buses 14–22. In other cases, these loads are supplied. In all the cases, most of the high significant loads are supplied such as buses 7–13, 24–25, and 26–29. In the 69-bus network, buses 48–50 and 61–65 supply high significant loads, and they are curtailed in case 1, while these buses are supplied under cases 2 and 3 where the proposed formation is implemented. Specially, buses 50 and 61 are

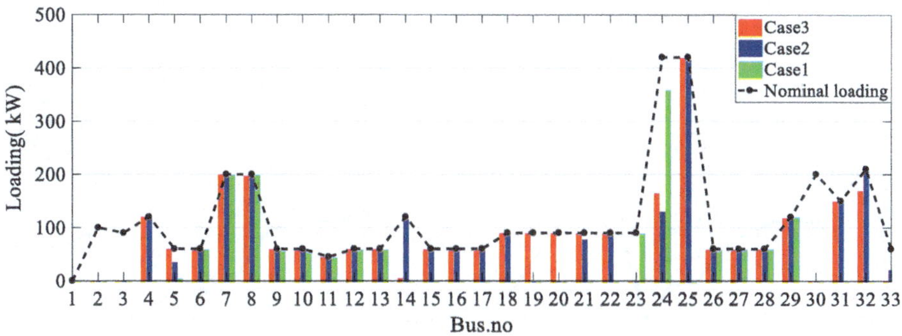

Fig. 6.12 Active loading of the 33-bus network under different cases

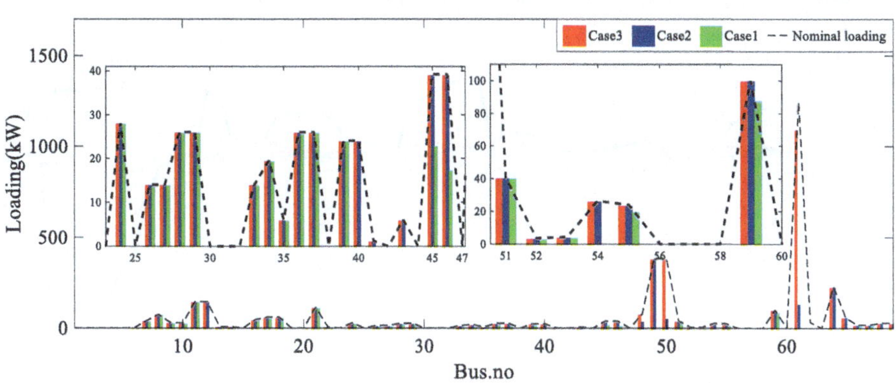

Fig. 6.13 Active loading of the 69-bus network under different cases

supplied under case 3, and this point shows the importance of tie lines in microgrid formation strategy.

6.6.2 Voltage Profile Assessment

The voltage amplitude in 33- and 69-bus networks under different cases is shown in Figs. 6.14 and 6.15. In case 3, the voltage alteration (deviation from 1 per unit) is lower than the other cases which means the tie lines can improve the voltage profile in the networks. In the 33-bus network, the voltage alteration in case 1 is lower than case 2 due to lower loading level (52%), and voltage amplitude on buses 1–4, 18–22, and 30–33 is high because in these buses the loads are curtailed while the solar PV and wind units are installed. The voltage deviation in the 69-bus network under case 1 is low, but this point is due to lower loading level (22%) as shown in Table 6.3. As an important result, the tie lines reduce the voltage alterations in both the 33- and 69-bus networks.

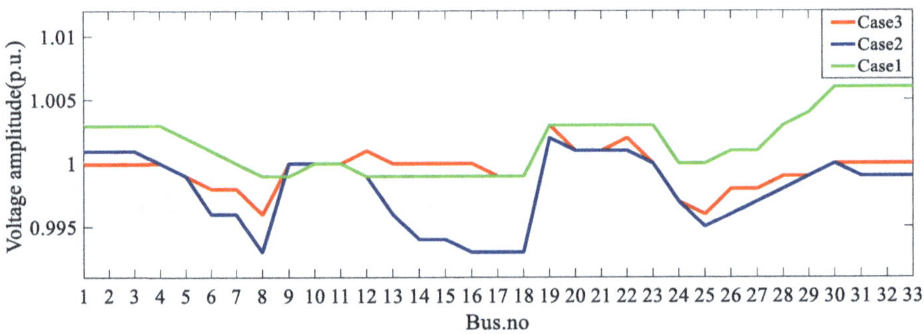

Fig. 6.14 Voltage amplitude of 33-bus network under different cases

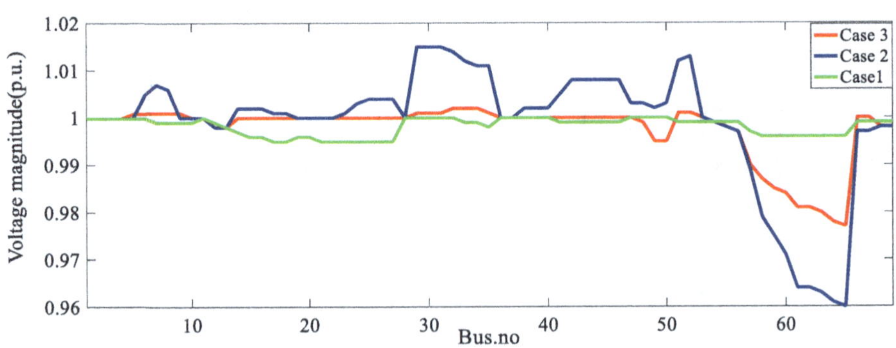

Fig. 6.15 Voltage amplitude of 69-bus network under different cases

6.6.3 DG Generation Evaluation

The active and reactive powers of DGs under different cases in 33- and 69-bus networks are presented in Tables 6.4 and 6.5. As seen in the tables, the tie lines almost have reduced the DGs' active power except some DGs such as the DG on bus 4 in the 33-bus network due to eliminating solar and wind units. As a result, the developed microgrid formation not only increases the load restoration but also reduces the DG generation, fuel cost, and environmental pollutions. In the proposed model, the costs have not been included and modelled. However, the microgrid formation approach not only reduces the penalty cost of load shedding through load restoration enhancement but also reduces DG fuel cost through reduction of DG generation. As a result, the total cost of system is reduced.

6.7 Conclusions

This chapter presents a microgrid formation strategy in order to maximize load restoration under faulty conditions. In the proposed method, the microgrid is islanded by opening breaker switches. The switch on-off states are optimized for load restoration maximization and blackout minimization. The created islands must satisfy the connectivity and radiality constraints. For this reason, a fictitious network is formed based on graph theory. The problem is formulated as MILP. In each case, it is assumed that the microgrid is isolated from the main grid. In the first case, the load restoration must be maximized without formation. In the second case, the microgrid is islanded with optimization of line breaker switch states. In the third case, the microgrid is islanded with both the line and tie breaker switches. Finally, the effectiveness of the proposed method is verified on modified IEEE 33-bus and 69-bus test systems. The simulation results show that the load restoration in 33-bus and 69-bus systems under case 1 is 52% and 22%, respectively. This index is 74% and 58% in 33-bus and 69-bus systems under case 2, and it is 89% and 95% under case 3. It is also shown that the microgrid formation not only improves load restoration but also reduces the power generation by DGs. It is verified that the

Table 6.4 Active and reactive powers of DG under different cases in the 33-bus network

	DG 4		DG 10		DG 23		DG 30	
	Active (kW)	Reactive (kVar)	Active (kW)	Reactive (kVar)	Active (kW)	Reactive (kVar)	Active (kW)	Reactive (kVar)
Case 1	371.46	148.36	382.79	113.86	371.46	148.36	371.46	148.36
Case 2	346.72	198.79	262.22	301.33	315.19	245.34	255	308
Case 3	383.169	112.71	46.92	300.81	315.19	245.34	92.65	388.35

Table 6.5 Active and reactive powers of DG under different cases in the 66-bus network

	DG 4		DG 11		DG 19		DG 22		DG 36	
	Active (kW)	Reactive (kVar)	Active (kW)	Reactive (kVar)	Active (kW)	Reactive (kVar)	Active (kW)	Reactive (kVar)	Active (kW)	Reactive (kVar)
Case 1	55.48	294.82	262.46	145.29	262.92	144.45	297.51	38.52	190.59	231.67
Case 2	16.1	130.89	220.17	203.58	93.97	284.82	156	256	3.5	229.87
Case 3	9.36	299.40	278.60	111.27	10	279.4	109.28	37.62	127	271

proposed model (case 3) including both the line and tie breaker switches presents the most optimal outputs.

References

1. Y. Wang, C. Chen, J. Wang, R. Baldick, Research on resilience of power systems under natural disasters—A review. IEEE Trans. Power Syst. **31**(2), 1604–1613 (2015)
2. A.M. Mohan, N. Meskin, H. Mehrjerdi, A comprehensive review of the cyber-attacks and cyber-security on load frequency control of power systems. *Energies* **13**(15), 3860 (2020)
3. R. Billinton, J. Billinton, Distribution system reliability indices. IEEE Trans. Power Deliv. **4**(1), 561–568 (1989)
4. I. t. c. K. Zetter, unprecedented hack of Ukraine's power grid, *Wired Magazine*, vol. 3, 2016
5. B. s. s. a. s. A. E. M. Operator, Report of the Australian Energy Market Operator Limited (AEMO), 2017
6. https://www.eia.gov/todayinenergy/detail.php?id=32992
7. A.O. Otuoze, M.W. Mustafa, R.M. Larik, Smart grids security challenges: Classification by sources of threats. J. Electr. Syst. Inf. Technol. **5**(3), 468–483 (2018)
8. M. Rahmani-Andebili, Analyzing the Effects of Problem Parameters on the Operation Cost of the Networked Microgrids, in *2020 IEEE Kansas Power and Energy Conference (KPEC)*, (2020) IEEE, pp. 1–6
9. W. Pimjaipong, T. Junrussameevilai, N. Maneerat, Blackout Prevention Plan-the Stability, Reliability and Security Enhancement in Thailand Power grid, in *2005 IEEE/PES Transmission & Distribution Conference & Exposition: Asia and Pacific*, (2005) IEEE, pp. 1–6
10. M.N. Ambia, K. Meng, W. Xiao, Z.Y. Dong, Nested formation approach for networked microgrid self-healing in islanded mode. IEEE Trans. Power Deliv. **36**, 452–464 (2020)
11. L. Che, M. Khodayar, M. Shahidehpour, Only connect: Microgrids for distribution system restoration. IEEE Power Energy Mag. **12**(1), 70–81 (2013)
12. H. Mehrjerdi, M. Saad, S. Lefebvre, Efficiency-resilience nexus in building energy management under disruptions and events. IEEE Syst. J. (2020)
13. C. Chen, J. Wang, F. Qiu, D. Zhao, Resilient distribution system by microgrids formation after natural disasters. IEEE Trans. Smart Grid **7**(2), 958–966 (2015)
14. I.-K. Song, W.-W. Jung, J.-Y. Kim, S.-Y. Yun, J.-H. Choi, S.-J. Ahn, Operation schemes of smart distribution networks with distributed energy resources for loss reduction and service restoration. IEEE Trans. Smart Grid **4**(1), 367–374 (2012)
15. C.-C. Liu, S.J. Lee, S. Venkata, An expert system operational aid for restoration and loss reduction of distribution systems. IEEE Trans. Power Syst. **3**(2), 619–626 (1988)
16. C.-S. Chen, C.-H. Lin, H.-Y. Tsai, A rule-based expert system with colored Petri net models for distribution system service restoration. IEEE Trans. Power Syst. **17**(4), 1073–1080 (2002)
17. S.-I. Lim, S.-J. Lee, M.-S. Choi, D.-J. Lim, B.-N. Ha, Service restoration methodology for multiple fault case in distribution systems. IEEE Trans. Power Syst. **21**(4), 1638–1644 (2006)
18. S.-J. Lee, S.-I. Lim, B.-S. Ahn, Service restoration of primary distribution systems based on fuzzy evaluation of multi-criteria. IEEE Trans. Power Syst. **13**(3), 1156–1163 (1998)
19. J.M. Solanki, S. Khushalani, N.N. Schulz, A multi-agent solution to distribution systems restoration. IEEE Trans. Power Syst. **22**(3), 1026–1034 (2007)
20. C.P. Nguyen, A.J. Flueck, Agent based restoration with distributed energy storage support in smart grids. IEEE Trans. Smart Grid **3**(2), 1029–1038 (2012)
21. A. Morelato, A. Monticelli, Heuristic search approach to distribution system restoration. IEEE Power Eng. Rev. **9**(10), 65–66 (1989)
22. S. Khushalani, J.M. Solanki, N.N. Schulz, Optimized restoration of unbalanced distribution systems. IEEE Trans. Power Syst. **22**(2), 624–630 (2007)

23. J. Zhu, Y. Yuan, W. Wang, An exact microgrid formation model for load restoration in resilient distribution system. Int. J. Electr. Power Energy Syst. **116**, 105568 (2020)
24. T. Ding, Y. Lin, Z. Bie, C. Chen, A resilient microgrid formation strategy for load restoration considering master-slave distributed generators and topology reconfiguration. Appl. Energy **199**, 205–216 (2017)
25. R. Hemmati, H. Saboori, P. Sianom, Coordinated short-term scheduling and long-term expansion planning in microgrids incorporating renewable energy resources and energy storage systems. Energy **134**, 699–708 (2017)
26. T. Ding, Y. Lin, G. Li, Z. Bie, A new model for resilient distribution systems by microgrids formation. IEEE Trans. Power Syst. **32**(5), 4145–4147 (2017)
27. K.S.A. Sedzro, X. Shi, A.J. Lamadrid, L.F. Zuluaga, A heuristic approach to the post-disturbance and stochastic pre-disturbance microgrid formation problem. IEEE Trans. Smart Grid **10**(5), 5574–5586 (2018)
28. K.S.A. Sedzro, A.J. Lamadrid, L.F. Zuluaga, Allocation of resources using a microgrid formation approach for resilient electric grids. IEEE Trans. Power Syst. **33**(3), 2633–2643 (2017)
29. M.A. Gilani, A. Kazemi, M. Ghasemi, Distribution system resilience enhancement by microgrid formation considering distributed energy resources. Energy **191**, 116442 (2020)
30. A. Kumar, S. Grijalva, Graph Theory and Critical Load-Based Distribution System Restoration using Optimal Microgrids Formation, in *2018 Clemson University Power Systems Conference (PSC)*, (2018) IEEE, pp. 1–6
31. S. Lei, C. Chen, Y. Song, Y. Hou, Radiality constraints for resilient reconfiguration of distribution systems: Formulation and application to microgrid formation. IEEE Transactions on Smart Grid (2020)
32. X. Wu, S. Shi, Z. Wang, Microgrid Planning Considering the Supply Adequacy of Critical Loads under the Uncertain Formation of Sub-Microgrids. *Sustainability* **11**(17), 4683 (2019)
33. H. Mehrjerdi, Resilience improvement with zero load curtailment by multi-microgrid based on system of systems. IEEE Access **8**, 198494–198502 (2020)
34. M.N.S. Noorpi, K. Meng, X. Li, Z.Y. Dong, W. Kong, Zonal Formation for Multiple Microgrids using Load Flow Sensitivity Analysis, in *2018 International Conference on Power System Technology (POWERCON)*, (2018) IEEE, pp. 358–363
35. H. Momen, A. Abessi, S. Jadid, Using EVs as distributed energy resources for critical load restoration in resilient power distribution systems. IET Gener. Trans. Distr. **14**(18), 3750–3761 (2020)
36. H. Mehrjerdi, Resilience-uncertainty nexus in building energy management integrated with solar system and battery storage, IEEE Access, (2020)
37. M. Rahmani-Andebili, Dynamic and adaptive reconfiguration of electrical distribution system including renewables applying stochastic model predictive control. IET Gener. Trans. Distrib. **11**(16), 3912–3921 (2017)
38. M. Rahmani-Andebili, Stochastic, adaptive, and dynamic control of energy storage systems integrated with renewable energy sources for power loss minimization. Renew. Energy **113**, 1462–1471 (2017)
39. M. Rahmani-Andebili, Cooperative distributed energy scheduling in microgrids, in *Electric Distribution Network Management and Control*, (Springer, Singapore, 2018), pp. 235–254
40. M. Rahmani-Andebili, Robust Operation of a Reconfigurable Electrical Distribution System by Optimal Charging Management of Plug-In Electric Vehicles Considering the Technical, Social, and Geographical Aspects, in *Planning and Operation of Plug-In Electric Vehicles*, (Springer, Cham, 2019), pp. 75–104
41. M. Rahmani-Andebili, M. Fotuhi-Firuzabad, An adaptive approach for PEVs charging management and reconfiguration of electrical distribution system penetrated by renewables. IEEE Trans. Ind. Inf. **14**(5), 2001–2010 (2017)

42. T. Ding, K. Sun, C. Huang, Z. Bie, F. Li, Mixed-integer linear programming-based splitting strategies for power system islanding operation considering network connectivity. IEEE Syst. J. **12**(1), 350–359 (2015)
43. R.A. Jabr, R. Singh, B.C. Pal, Minimum loss network reconfiguration using mixed-integer convex programming. IEEE Trans. Power Syst. **27**(2), 1106–1115 (2012)
44. R. Balakrishnan, K. Ranganathan, *A Textbook of Graph Theory* (Springer, New York, 2012)
45. H. Zhang, V. Vittal, G.T. Heydt, J. Quintero, A mixed-integer linear programming approach for multi-stage security-constrained transmission expansion planning. IEEE Trans. Power Syst. **27** (2), 1125–1133 (2011)
46. H. Mehrjerdi, R. Hemmati, Modeling and optimal scheduling of battery energy storage systems in electric power distribution networks. J. Clean. Prod. **234**, 810–821 (2019)
47. H. Mehrjerdi, Simultaneous load leveling and voltage profile improvement in distribution networks by optimal battery storage planning. Energy **181**, 916–926 (2019)
48. H. Saboori, S. Jadid, Optimal scheduling of mobile utility-scale battery energy storage systems in electric power distribution networks. J. Energy Storage **31**, 101615 (2020)
49. R. Parasher, Load flow analysis of radial distribution network using linear data structure, *arXiv preprint arXiv:1403.4702,* 2014
50. I. L. O. G. U. s. m. I. S. Cplex, Gentilly, France (2007): 32
51. A. Brook, D. Kendrick, A. Meeraus, GAMS, a user's guide. ACM Signum Newsl. **23**(3–4), 10–11 (1988)

Index

M. Rahmani-Andebili (ed.), *Design, Control, and Operation of Microgrids in Smart Grids*, Power Systems, https://doi.org/10.1007/978-3-030-64631-8

The manufacturer's authorised representative in the EU is Springer Nature Customer Service Centre GmbH, Europaplatz 3, 69115 Heidelberg, Germany. If you have any concerns regarding our products, please contact ProductSafety@springernature.com

Printed and bound by CPI Group (UK) Ltd, Croydon, CR0 4YY
28/09/2026
02224090-0001